Expand Your Math Power!

Pre-Algebra

- Quickly grasp key concepts!

- Practice with hundreds of Problems!

- Achieve higher math proficiency quickly!

Contents

Welcome Students!

You did the right choice in deciding to work with this guide. Your time is very valuable and I don't want you to waste it in long exercises and tedious explanations that do not add any meaningful understanding to the concepts and skills required to master pre-algebra.

Instead, this guide is just straight to the point! Each and almost every page starts with quick and short example or explanation immediately followed by the skill practice. Work on these exercises and check the selected answers (usually odd numbers) at the back of the book! That is it. In no time, you master the pre-algebra concepts and be ready to cruise algebra at high speed!

Congratulations in deciding to master the pre-algebra which is the true entrance to advanced algebra!

Please do not hesitate to write me any comments or questions about this book at aalbotan@yahoo.com.

1) Find the Primes

A Prime number is a whole number that can be divided evenly by 1 and itself only. For example 7 can be divided evenly by 1 or 7 only.

But, **8** can divide evenly by1, 2, 4 and itself. So, it is not prime number. It is a **composite.**

Circle the primes in each question

1)   3,   4,   5,   7,   9,   11,   13,   15,   17

2)   18,   19,   21,   23,   27,   29,   31,   33,   37,   39

3)   41,   43,   45,   46,   47,   48,   49,   51,   52,   53

4)   55,   57,   58,   59,   60,   61,   62,   63,   64,   65

5)   66,   67,   69,   71,   73,   75,   77,   78,   79,   81

6)   83,   85,   87,   89,   91,   93,   96,   97,   99,   101

2) Prime Factorization

Example: Find the prime factorization of: (a) 84;   (b) 138

Divide the number continuously until you have no other number to divide to it except to itself and 1!

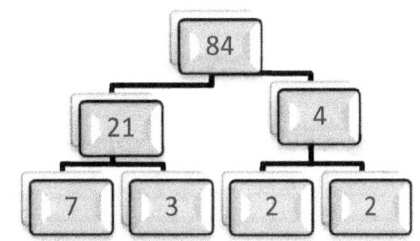

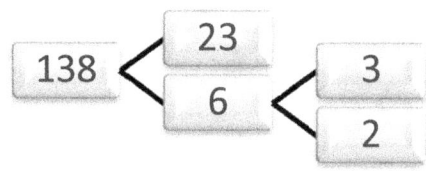

(a) The answer: $2^2 \times 3 \times 7$ **(b) The answer:** $2 \times 3 \times 23$

1) 16   2) 48   3) 56

4) 63   5) 39   6) 112

7) 100   8) 96   9) 75

10) 325   11) 128   12) 396

3) Factoring

To factor a number means to break the number as a product of other numbers. The way to find the factor is to find which numbers that can divide evenly the number.

Example 1: **Find the factors of 40**	**Example 2:** **List all the factors of 42**
$40 \div 1 = 40$	:
$40 \div 2 = 20$	$42 \div 1 = 42$
$40 \div 4 = 10$	$42 \div 2 = 21$
$40 \div 5 = 8$	$42 \div 3 = 14$
	$42 \div 6 = 7$
So, the factors are:1, 2, 4, 5, 8, 10,20,40	So, the factors are: 1, 2, 3, 6, 7, 14, 21, and42.
All these numbers divide 40 evenly.	*All these numbers divide 42 evenly.*

1) 16

2) 48

3) 56

4) 63

5) 84

6) 25

7) 136

8) 105

9) 35

10) 117

11) 100

12) 75

4) Find Greatest Common Factors (GCF): Use Factor Lists

The name says all! To find the greatest factor (GCF) of two numbers, (a) First, list the factors of each number and then (2) choose the greatest common number!

(14, 28)	**14:**	1	2	**7**	(14)	
	28:	1	2	4	7	(14)	28

See the common factors are 1, 2, 7, and 14. But we choose the greatest which is **14.**

		List the Factors here	The GCDF is
1)	**(6,8)**	6: 1 **2** 3 6	**2**
		8: 1 **2** 4 8	
2)	**(12,16)**	12:	
		16	
3)	**(7,28)**	7:	
		28:	
4)	**(8,14)**	8:	
		14:	
5)	**(11, 22)**	11:	
		22:	
6)	**(16, 48)**	16:	
		48:	
7)	**(9,16)**	9:	
		16:	
8)	**(12,24)**	12:	
		24:	
9)	**(10, 20)**	10:	
		20:	
10)	**(6, 24)**	6:	
		24:	

5) Greatest Common Factors (GCF): Use Division Ladder

	EXAMPLE A) **4 and 10**	EXAMPLE B) **9 and 12**
✓ Continue to divide (the 4, and 10) or the (14 and 21) into numbers such as 2, 3, 7 etc. Until you can't find common number between numbers! ✓ Multiply the first raw numbers	<table><tr><td></td><td>**4**</td><td>**10**</td></tr><tr><td>**÷2**</td><td>2</td><td>5</td></tr><tr><td></td><td></td><td></td></tr></table> **Since the 2 and 5 have no common number:** **The GCF = 2**	<table><tr><td></td><td>**14**</td><td>**21**</td></tr><tr><td>**÷7**</td><td>2</td><td>3</td></tr><tr><td></td><td></td><td></td></tr></table> **Since the 2 and 3 have no common number:** **GCF = 7**

1. $(7, 21)$

2. $(12, 10)$

3. $(9, 18)$

4. $(15, 5)$

5. $(14, 8)$

6. $(20, 15)$

7. $(24, 16)$

8. $(12, 18)$

9. $(24, 18)$

10. $(6, 18)$

11. $(10, 30)$

12. $(18, 30)$

6) Find Least Common Multiple (LCM): Use Factor Lists

The name says all! To find the lowest common multiple, just continuously multiply 1, 2, 3 etc. to each number till you get a common multiple:

14   14   **28**   **See. This time we quickly got a match of 28. So, LCM= 28**

28   **28**   56

Write the multiples, then circle the least common multiple for each pair of numbers:

		List the multiples here	The LCM is
1)	**(6,8)**	6: 6 12 18 **24**	**24**
		8: 8 16 **24** 36	
2)	**(12,16)**	12:	
		16	
3)	**(7,28)**	7:	
		28:	
4)	**(8,14)**	8:	
		14:	
5)	**(11, 22)**	11:	
		22:	
6)	**(16, 48)**	16:	
		48:	
7)	**(9,16)**	9:	
		16:	
8)	**(12,24)**	12:	
		24:	
9)	**(10, 20)**	10:	
		20:	
10)	**(6, 24)**	6:	
		24:	

7) The Least Common Multiple (LCM): Use Division Ladder

	EXAMPLE A) *4 and 10*		EXAMPLE B) *9 and 12*	
✓ Continue to divide the 4, and 10 or the 9 and 12 into numbers such as 2, 3, 5 etc. Until all bottom numbers turn into ones!		**4** / **10**		**9** / **12**
	÷2	2 / 5	÷3	3 / 4
	÷2	1 / 5	÷3	1 / 4
	÷5	/ 1	÷4	/ 1
✓ If the number is not divisible just copy it again until you divide it. ✓ Multiply the first raw numbers.	LCM = $2 \times 2 \times 5 = 20$		LCM = $3 \times 3 \times 4 = 36$	

1. $(7, 14)$

2. $(6, 10)$

3. $(9, 18)$

4. $(15, 5)$

5. $(14, 8)$

6. $(20, 10)$

7. $(20, 16)$

8. $(12, 18)$

9. $(24, 18)$

10. $(6, 18)$

11. $(10, 30)$

12. $(18, 30)$

8) Review of Fractions: Improper & Mixed

1) Change the improper fraction into mixed numbers:	$\dfrac{13}{3}$	$3\overline{)13}$ 4 $\underline{12}$ 1	$\dfrac{13}{3} = 4\dfrac{1}{3}$

1) $\dfrac{7}{4}$ 6) $\dfrac{13}{3}$ 11) $\dfrac{23}{6}$

2) $\dfrac{51}{2}$ 7) $\dfrac{17}{2}$ 12) $\dfrac{69}{4}$

3) $\dfrac{17}{3}$ 8) $\dfrac{51}{5}$ 13) $\dfrac{87}{7}$

4) $\dfrac{49}{8}$ 9) $\dfrac{75}{6}$ 14) $\dfrac{96}{9}$

5) $\dfrac{27}{8}$ 10) $\dfrac{93}{8}$ 15) $\dfrac{33}{5}$

2) Change the mixed numbers into improper fraction:	$7\dfrac{3}{4}$	$4 \times 7 + 3 = 31$	$7\dfrac{3}{4} = \dfrac{31}{4}$

16) $3\dfrac{1}{4}$ 17) $1\dfrac{3}{13}$ 18) $10\dfrac{1}{2}$

19) $6\dfrac{1}{2}$ 20) $4\dfrac{4}{5}$ 21) $11\dfrac{3}{4}$

22) $8\dfrac{2}{3}$ 23) $7\dfrac{3}{5}$ 24) $12\dfrac{2}{7}$

25) $5\dfrac{4}{5}$ 26) $8\dfrac{4}{9}$ 27) $13\dfrac{3}{5}$

9) Review of Fractions: Simplify

3) **Simplify Fractions:**	$\dfrac{16}{24}$	*Divide **both** the denominator and numerator by the greatest common factor(GCF):*	$\dfrac{16 \div 8}{24 \div 8} = \dfrac{2}{3}$

1) $\dfrac{2}{8}$ 6) $\dfrac{6}{16}$ 11) $\dfrac{25}{40}$

2) $\dfrac{3}{9}$ 7) $\dfrac{7}{21}$ 12) $\dfrac{35}{50}$

3) $\dfrac{5}{10}$ 8) $\dfrac{8}{24}$ 13) $\dfrac{40}{50}$

4) $\dfrac{4}{12}$ 9) $\dfrac{9}{36}$ 14) $\dfrac{65}{15}$

5) $\dfrac{5}{15}$ 10) $\dfrac{10}{20}$ 15) $\dfrac{33}{44}$

4) **Simplify Mixed Numbers**	$8\dfrac{6}{24}$	*Divide **both** the denominator and numerator of the fraction part by the greatest common factor(GCF):*	$8\dfrac{6 \div 6}{24 \div 6} = 8\dfrac{1}{4}$

16) $3\dfrac{2}{4}$ 20) $1\dfrac{3}{12}$ 24) $10\dfrac{8}{22}$

17) $7\dfrac{6}{10}$ 21) $4\dfrac{4}{20}$ 25) $11\dfrac{12}{16}$

18) $4\dfrac{2}{8}$ 22) $7\dfrac{5}{25}$ 26) $12\dfrac{14}{28}$

19) $5\dfrac{6}{9}$ 23) $8\dfrac{4}{10}$ 27) $13\dfrac{30}{50}$

10) Fractions: Add/Subtract Like Denominators

	Example 1	Example 2
1. **Add** the whole Numbers. 2. **Add/subtract** the numerators 3. **Keep** the denominator unchanged. 4. **Simplify/rename** as needed.	$\dfrac{3}{8} + \dfrac{1}{8}$ $= \dfrac{3+1}{8} = \dfrac{4}{8}$ $\dfrac{4 \div 4}{8 \div 4} = \dfrac{1}{2}$	$3\dfrac{3}{10} + 1\dfrac{2}{10}$ $= 4\dfrac{5}{10}$ $4\dfrac{5 \div 5}{10 \div 5} = 4\dfrac{1}{2}$

1) $\dfrac{1}{5} + \dfrac{2}{5}$

2) $\dfrac{2}{7} + \dfrac{3}{7}$

3) $\dfrac{3}{8} + \dfrac{1}{8}$

4) $\dfrac{5}{9} - \dfrac{4}{9}$

5) $\dfrac{5}{12} + \dfrac{1}{12}$

6) $6\dfrac{3}{5} - 3\dfrac{1}{5}$

7) $1\dfrac{1}{5} + 3\dfrac{2}{5}$

8) $5\dfrac{1}{4} + 2\dfrac{1}{4}$

9) $2\dfrac{2}{10} + 7\dfrac{1}{10}$

10) $6\dfrac{2}{5} - 5\dfrac{1}{5}$

11) $\dfrac{9}{21} - \dfrac{2}{21}$

12) $\dfrac{3}{11} + \dfrac{5}{11}$

13) $\dfrac{3}{10} + \dfrac{6}{10}$

14) $\dfrac{4}{15} + \dfrac{3}{15}$

15) $\dfrac{9}{14} - \dfrac{5}{14}$

16) $7\dfrac{5}{12} + 12\dfrac{1}{12}$

17) $4\dfrac{4}{20} - 3\dfrac{3}{20}$

18) $7\dfrac{2}{25} + 8\dfrac{3}{25}$

19) $4\dfrac{15}{16} - 1\dfrac{9}{16}$

20) $8\dfrac{3}{10} - 1\dfrac{1}{10}$

11) Fractions: Add/Subtract Unlike Denominators

Examples: $\dfrac{1}{4} + \dfrac{1}{3}$ $2\dfrac{1}{2} + 1\dfrac{1}{5}$

1. Find the LCM. 2. Make the denominators equal to the LCM by multiplying right numbers. 3. Add/ Subtract and Simplify	*The LCM is 12* $= \dfrac{\mathbf{1} \times ③}{\mathbf{4} \times ③} + \dfrac{\mathbf{1} \times ④}{\mathbf{3} \times ④}$ $\dfrac{3}{12} + \dfrac{4}{12} = \dfrac{7}{12}$	*The LCM is 10* $2\dfrac{\mathbf{1} \times ⑤}{\mathbf{2} \times ⑤} + 1\dfrac{\mathbf{1} \times ②}{\mathbf{5} \times ②}$ $2\dfrac{5}{10} + 1\dfrac{2}{10} = 3\dfrac{7}{10}$

1) $\dfrac{1}{2} + \dfrac{7}{8}$

2) $\dfrac{3}{4} - \dfrac{1}{5}$

3) $\dfrac{2}{5} + \dfrac{1}{4}$

4) $\dfrac{2}{4} - \dfrac{2}{8}$

5) $\dfrac{1}{4} - \dfrac{1}{12}$

6) $\dfrac{2}{3} + \dfrac{2}{9}$

7) $\dfrac{1}{4} + \dfrac{2}{3}$

8) $\dfrac{5}{8} - \dfrac{1}{4}$

9) $\dfrac{4}{15} + \dfrac{7}{10}$

10) $2\dfrac{1}{2} + 1\dfrac{1}{4}$

11) $5\dfrac{2}{4} - 2\dfrac{1}{12}$

12) $4\dfrac{1}{5} + 3\dfrac{1}{3}$

13) $7\dfrac{2}{3} - 2\dfrac{2}{9}$

14) $1\dfrac{5}{9} + 1\dfrac{2}{5}$

15) $7\dfrac{3}{5} - \dfrac{2}{7}$

16) $4\dfrac{4}{5} - 3\dfrac{3}{20}$

17) $7\dfrac{2}{3} + 5\dfrac{3}{8}$

18) $4\dfrac{5}{12} - \dfrac{3}{14}$

12) Multiply Fractions & Mixed Numbers

	Example 1	Example 2
1. **Change** any mixed number into improper fractions. 2. **Multiply** the numerators and denominators. (Use cross cancelations if possible). **3. Simplify**	$\dfrac{3}{8} \times \dfrac{2}{5}$ $= \dfrac{6}{40} = \dfrac{3}{20}$	$2\dfrac{1}{10} \times 1\dfrac{3}{7}$ $= \dfrac{21^3}{10_1} \times \dfrac{10^1}{7^1} = \dfrac{3}{1} = 3$

1) $\dfrac{1}{5} \times \dfrac{2}{5}$

2) $\dfrac{2}{7} \times \dfrac{3}{4}$

3) $\dfrac{3}{5} \times \dfrac{5}{8}$

4) $\dfrac{5}{6} \times \dfrac{4}{15}$

5) $\dfrac{2}{21} \times \dfrac{7}{10}$

6) $\dfrac{3}{10} \times \dfrac{4}{3}$

7) $\dfrac{3}{14} \times \dfrac{7}{18}$

8) $\dfrac{5}{6} \times \dfrac{3}{20}$

9) $\dfrac{3}{8} \times \dfrac{7}{4}$

10) $1\dfrac{1}{5} \times 1\dfrac{1}{4}$

11) $5\dfrac{1}{5} \times 2\dfrac{1}{12}$

12) $4\dfrac{1}{5} \times 3\dfrac{1}{3}$

13) $2\dfrac{2}{9} \times 2\dfrac{7}{10}$

14) $1\dfrac{5}{9} \times 1\dfrac{2}{7}$

15) $3\dfrac{1}{2} \times 2\dfrac{4}{7}$

16) $4\dfrac{1}{5} \times 3\dfrac{4}{7} =$

17) $3\dfrac{3}{16} \times 1\dfrac{1}{17}$

18) $3\dfrac{1}{5} \times 3\dfrac{3}{4}$

13) Divide Fractions/Mixed Numbers

Example: Simplify

$$1\frac{1}{5} \div 2\frac{2}{5}$$

1. Rename any mixed numbers
2. Change the division sign into multiplication sign
3. Flip the divisor (the second fraction) & simplify

$$\frac{6}{5} \div \frac{12}{5}$$

$$\frac{6}{5} \times \frac{5}{12} = \frac{30 \div 30}{60 \div 30} = \frac{1}{2}$$

When it is possible, cross cancelation is easier

$$\frac{6^1}{5_1} \times \frac{5^1}{12^2} = \frac{1}{2}$$

1) $\dfrac{1}{5} \div \dfrac{5}{2}$

2) $\dfrac{4}{5} \div \dfrac{8}{5}$

3) $\dfrac{2}{7} \div \dfrac{3}{14}$

4) $\dfrac{5}{9} \div \dfrac{5}{18}$

5) $\dfrac{1}{2} \div \dfrac{5}{2}$

6) $\dfrac{3}{4} \div \dfrac{9}{4}$

7) $\dfrac{9}{8} \div \dfrac{9}{8}$

8) $\dfrac{7}{25} \div \dfrac{2}{5}$

9) $\dfrac{5}{9} \div \dfrac{3}{7}$

10) $\dfrac{9}{11} \div \dfrac{3}{22}$

11) $5\dfrac{2}{3} \div 4\dfrac{6}{7}$

12) $6\dfrac{3}{4} \div 3\dfrac{3}{4}$

13) $4\dfrac{1}{5} \div 1\dfrac{1}{6}$

14) $9\dfrac{1}{8} \div 2\dfrac{5}{6}$

15) $1\dfrac{1}{5} \div 3\dfrac{1}{4}$

16) $1\dfrac{1}{4} \div 3\dfrac{2}{3}$

17) $2\dfrac{4}{15} \div 2\dfrac{7}{10}$

18) $7\dfrac{2}{3} \div 7\dfrac{3}{4}$

19) $4\dfrac{1}{2} \div 1\dfrac{4}{5}$

20) $7\dfrac{1}{3} \div 4\dfrac{2}{5}$

14) What is Percent?

The word percent means "out of a hundred" or a part of a hundred. For example if you split with your friend $100, each takes $50 out of the $100 or **50 percent**. Instead of writing every time "Percent" we simply use the percent sign %.
So, 50 percent is written as 50 %.

Common Sense Percent Problems:

Write the following as a percent. Use common sense and your knowledge of fraction and the example above:

1) You split with three other friends $100. What percent will each get?

2) In every four free throws, Shadiya makes three. What percent Shadiya scores?

3) Ten brothers and sisters shared $800. What percent will each get?

4) Awil was the fourth in the math tests in a class of 10 at Pre-University. What percent of students were lower? What percent were higher?

From problems 5 to 7 use the pie chart:

5) Which two activities does Dalmar spends 50% of his time?

6) Estimate the percent of time Dalmar spends on playing.

7) What are the total percent of time Dalmar spends on all four subjects: Writing, reading, math and science?

8) Ouch! 75% of all accidents happen within 5 miles of home. Could you explain what does that mean?

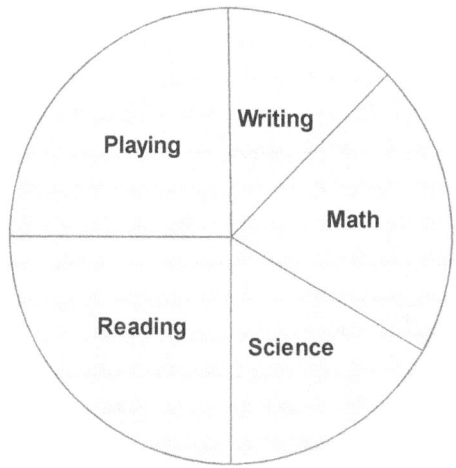

Dalmar's Activity Chart

15) Change Fractions into Percent

To change fractions into percent: Multiply it by 100 and add the % sign.	(a) $\frac{2}{5}$ $\frac{2}{5} \times 100\%$ $= \frac{2 \times 100}{5} = 40\%$	(b) $\frac{1}{4}$ $\frac{1}{4} \times 100$ $\frac{1 \times 100}{4} = 25\%$

Change the following fractions into percent:

1) $\frac{3}{4} =$

2) $\frac{4}{5} =$

3) $\frac{1}{4} =$

4) $\frac{3}{5} =$

5) $\frac{1}{5} =$

6) $\frac{4}{5} =$

7) $\frac{5}{6} =$

8) $\frac{1}{20} =$

9) $\frac{1}{12} =$

10) $\frac{1}{14} =$

11) $\frac{5}{12} =$

12) $\frac{1}{8} =$

13) $\frac{1}{24} =$

14) $\frac{3}{50} =$

15) $\frac{1}{60} =$

16) $\frac{2}{25} =$

16) Change Mixed Numbers into Percent: Method 1

1) Change the mixed numbers into fractions	$2\frac{1}{2} = \frac{5}{2}$	
2) Multiply it by 100%	$\frac{5}{2} \times 100\% = \frac{500}{2}$	
3) Simplify the fractions	$\frac{500}{2} = 250\%$	

1) $3\frac{1}{2} =$

2) $5\frac{1}{2} =$

3) $7\frac{1}{4} =$

4) $8\frac{3}{4} =$

5) $9\frac{1}{5} =$

6) $5\frac{2}{5} =$

7) $7\frac{4}{5} =$

8) $14\frac{1}{2} =$

9) $11\frac{1}{4} =$

10) $2\frac{1}{20} =$

11) $12\frac{3}{10} =$

12) $17\frac{1}{100} =$

17) Change Mixed Numbers into Percent: Method 2

1) Change the mixed numbers into decimals	$2\frac{1}{2} = 2 + 0.5 = 2.5$
2) Multiply it by 100%	$2.5 \times 100 = 250\%$

1) $3\frac{1}{2} =$

2) $5\frac{1}{2} =$

3) $7\frac{1}{4} =$

4) $8\frac{1}{2} =$

5) $8\frac{3}{4} =$

6) $9\frac{1}{5} =$

7) $5\frac{2}{5} =$

8) $12\frac{3}{5} =$

9) $7\frac{4}{5} =$

10) $14\frac{1}{2} =$

11) $11\frac{1}{4} =$

12) $3\frac{8}{10} =$

13) $2\frac{1}{20} =$

14) $12\frac{3}{10} =$

15) $17\frac{1}{100} =$

16) $9\frac{1}{80} =$

18) Change Percent into Fractions

Example: Change the percent into fraction: (a) 3%,   b) 12%

1) Just divide it by 100 & get rid of the percent sign	$\dfrac{3}{100}$	$\dfrac{12}{100}$
2) Simplify as needed	$\dfrac{3}{100}$	$\dfrac{12 \div 4}{100 \div 4} = \dfrac{3}{25}$

1. 13% =

2) 40% =

3) 17% =

4) 45% =

5) 200% =

6) 41% =

7) 24% =

8) 16% =

9) 55 % =

10) 11% =

11) 42% =

12) 28% =

13) 4% =

14) 54% =

15) 28% =

19) Change Mixed Number Percent into Fractions

Example: Change the percent into fraction: $2\frac{1}{2}\%$

1) **Change** any mixed numbers into fractions	$\dfrac{5}{2}$
2) **Multiply** the lower number by 100	$= \dfrac{5}{2 \times 100} = \dfrac{5}{200}$
3) **Simplify** the fractions	$= \dfrac{5 \div 5}{200 \div 5} = \dfrac{1}{40}$

1. $2\frac{1}{4}\% =$

2. $4\frac{1}{2}\% =$

3. $5\frac{1}{4}\% =$

4. $12\frac{1}{5}\% =$

5. $1\frac{3}{5}\% =$

6. $6\frac{2}{5}\% =$

7. $8\frac{1}{4}\% =$

8. $8\frac{1}{2}\% =$

9. $12\frac{2}{5}\% =$

10. $8\frac{3}{4}\% =$

11. $3\frac{1}{5}\% =$

12. $15\frac{2}{5}\% =$

20) Change Decimals into Percent

To change fractions into percent:	(a) 0.5	(b) 1.06
	↓	↓
Multiply it by 100 and add the % sign.	$0.5 \times 100\% =$ 50%	106%

1) 1.3

2) 2.3

3) 1.02

4) 4.5

5) 6.5

6) 4.0

7) 7.05

8) 8.25

9) 9.5

10) 0.05

11) 0.174

12) 14.1

13) 4.33

14) 3.12

15) 11.5

21) Change Percent into Decimals

Examples: Change the percent into Decimals: (a) $12\frac{1}{2}\%$ (b) 3%

Step by Step	(a) $12\frac{1}{2}\%$	(b) 3%
1. Change any mixed numbers into improper fraction.	$\frac{25}{2}\%$	3%
2. Multiply it by 0.01 and remove the % sign.	$\frac{25}{2} \times 0.01$ $= 0.125$	3×0.01 $= 0.03$

1) 13%   2) 40%   3) 17%   4) 2.5%

5) 1.3%   6) 45%   7) 200%   8) 0.2%

9) $4\frac{1}{2}\%$ 10) 0.75%   11) 40 %   12) 0.02

13) $5\frac{1}{2}\%$ 14) 0.25%   15) $3\frac{1}{5}\%$ 16) 0.012%

 *Did you know that the human brain is 80% water?*

22) Change Decimal Percent into Fractions

Example1: Change the percent into fraction: (a) 0.5% b) 0.24%

	0.5%	0.24%
1) Divide it by 100. Rid of % sigh	$\dfrac{0.5}{100}$	$\dfrac{0.24}{100}$
2) Get rid of the decimal. Multiply & divide with 10, 100,1000...as needed	$\dfrac{0.5}{100} \times \dfrac{10}{10} = \dfrac{5}{1000}$	$\dfrac{0.24}{100} \times \dfrac{100}{100} = \dfrac{24}{10000}$
3) Simplify the fractions	$= \dfrac{5}{1000} = \dfrac{1}{200}$	$\dfrac{24}{1000} = \dfrac{3}{125}$

1) 1.3% =   2) 0.4% =   3) 1.7% =

4) 4.5% =   5) 2.0% =   6) 4.1% =

7) 0.22% =   8) 1.6% =   9) 5.5 % =

10) 1.1% =   11) 4.5% =   12) 2. 8% =

 **Scary Percent:** According to a study reported in Washington Post in 2010, 28% of traffic accidents occur when people talk on cellphones or send text messages while driving!

23) Mixed Review: Fractions, Decimals, Percent

Fill in the blanks with the correct fraction, percent or decimal:

	Fraction	Decimal	Percent
1)	$\dfrac{2}{5}$	0.4	40%
2)	$\dfrac{1}{5}$		
3)		0.7	
4)			12%
5)	$\dfrac{3}{4}$		
6)		21.4	
7)			1.5%
8)	$\dfrac{5}{8}$		
9)		5.4	
10)	$12\dfrac{1}{5}$		
11)			13.8%
12)	$12\dfrac{1}{5}$		
13)			0.06%

24) Percent: Finding the Part

> **Example: What is 6% of 300?**
>
> $Part = Percent \ (in \ decimal) \times Base$ $\qquad$ $Part = 0.06 \times 300 = 50$

1) What is 2% of 200?

2) What is 4% of 300?

3) What is 15% of 500?

4) What is 0.25% of 62?

5) What is 8.2 % of 500?

6) What is 7% of 6.54?

7) What is 21% of 800?

8) What is 22% of 1600?

9) What is 0.05% of 1000?

10) What is 12% of 700?

25) Finding the Percent

What percent is \$10 out of \$120?	(Or \$10 is what percent of \$120?)
$Percent = \dfrac{Part}{Base} \times 100$ $Percent = \dfrac{10}{120} \times 100 = 50\%$	

1) What percent is \$2 out of \$50?

2) \$4 is what percent of \$56?

3) 2 is what percent of 8?

4) What percent is 4 out of 60?

5) What percent is 24 out of 504?

6) What percent of 400 is 80?

7) What percent is 20 out of 800?

8) \$70 is what percent \$560?

9) What percent of 350 is 7?

10) What percent of 12 is 4?

26) Finding the base

25 % of what number is 40? *You can do it in two ways:*	
$Base = \dfrac{Amount}{Percent\ (in\ decimals)}$ $= \dfrac{40}{0.25} = 160$	$Base = \dfrac{Amount}{Percent} \times 100$ $= \dfrac{40}{25} \times 100 = 160$

1) 18% of what number is 54?

2) 50% of what number is 1.5?

3) 60 is 20% of what number?

4) 2.8 is 50% of what number?

5) 2% of what number is 10?

6) 95 is 45% of what number?

7) 5% of what number is 50?

8) 280 is 14% of what number?

9) 10% of what number is 47?

10) 210% of what number is 525?

27) Percent: Mixed Practice

1) What is 5% of 200?

2) 4% of what number is 30?

3) $20 is what percent of 500?

4) What is 25% of 800?

5) 75 is what percent of 600?

6) What is 13% of 90?

7) What percent of 620 is 248?

8) What is 12% of 1600?

9) 15% of what number is 12?

10) What is 0.052% of 8100?

28) Find the Percent Change

Example: The price of a shirt decreased from $20 to $14. What is the percent of change?

$$\textbf{Percent Change} = \frac{|Original\ amount - new\ amount|}{Original} \times \textbf{100\%}$$

$$Percent\ Change = \frac{|20 - 14|}{20} \times 100\% = \frac{6}{20} \times 100\% = 30\%$$

Important: the little bars | | in the formula is the absolute value. That means the answer should always be positive!

1) The price of math books increased from $10 to $12 this week end. What is the percent change?

2) On the New Year's Eve the price of shoes decreased from $30 to $20. What is the percent change?

3) Kadar's monthly income for the last year was $2500.This year it is $2900. What is his percentage increase?

4) The speed of car increased from 40 mph to 55 mph. What is the percent increase?

5) Last year, the total number of students of the new school was 200. This year it has increased to 350. What is the percent increase?

6) Siman was 66 inches tall before her height increased by 4 inches. What is the percent change in her height?

7) Farah's weight on January was 152 pounds. This month he lost 12 pounds. What is the percent change?

8) Ahmed practices math for 12 hours a week. Now he is planning to increase it to 16 hours. What will be the percent change?

29) Find the New Amount: Tax, Discount etc.

Example 1: How much you will pay for a shirt costing $40 with 10% tax? $40 + 0.10 ($40) = $44	**Example 2**: The price of a shirt was $20 but is now decreased by 15%. What is the price now? $20 - 0.15 ($20) = $17

1) What is $400 increased by 20%?

2) In December, the price of shoes decreased by 25%. The original price was $36 What is the new price?

3) Tony's monthly income for last year was $2500 before it increased by 10%. What is his new monthly income?

4) The cost of a car was $15,000 before adding 8% tax. What is the total cost?

5) The number of students for the new school was 300. This year it has increased by 12%. How many students are there now?

6) Sooyan was 60 inches tall before her height increased by 4% inches. What is his new height?

7) Leyla's weight on January was 152 pounds. This month she lost 12% of her weight. What is her weight now?

8) Awale practices biology for 12 hours a week. Now he is planning to increase it by 18%. What will be the new number of hours for practice?

30) Review of Perfect Squares

If you know these squares on the top of your head, you will have easy time with the exponents and square roots!

Exercise 1: Simplify the following squares;

$0^2 =$ $1^2 =$ $2^2 =$ $3^2 =$

$4^2 =$ $5^2 =$ $6^2 =$ $7^2 =$

$8^2 =$ $9^2 =$ $10^2 =$ $11^2 =$

$12^2 =$ $13^2 =$ $14^2 =$ $15^2 =$

$16^2 =$ $17^2 =$ $18^2 =$ $19^2 =$

$20^2 =$ $21^2 =$ $22^2 =$ $23^2 =$

$24^2 =$ $25^2 =$ $26^2 =$ $27^2 =$

It is worth to memorize these perfect squares.

31) Square Roots

Square roots are the inverse of the squares

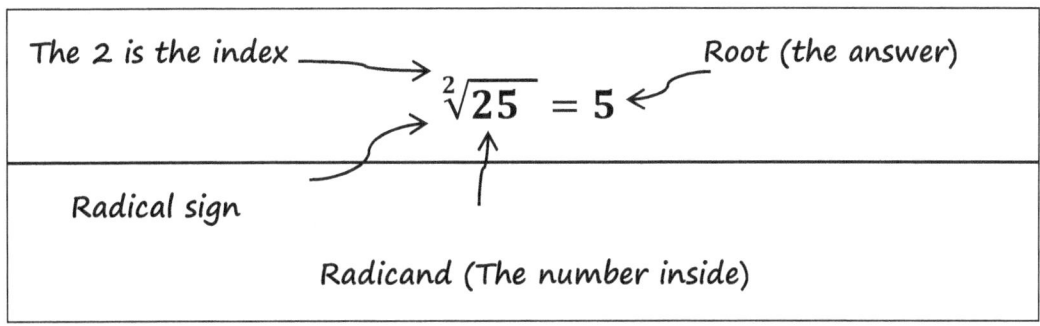

Since $5^2 = 25$, the square root of 25 is 5

Note: The square root is so common that we usually omit the indices or the 2 and write simply it as: $\sqrt{25}$

Find the square roots of the following:

$\sqrt{16}$	$\sqrt{49}$	$\sqrt{1}$	$\sqrt{36}$
$\sqrt{9}$	$\sqrt{4}$	$\sqrt{0}$	$\sqrt{64}$
$\sqrt{81}$	$\sqrt{100}$	$\sqrt{169}$	$\sqrt{121}$
$\sqrt{400}$	$\sqrt{256}$	$\sqrt{144}$	$\sqrt{196}$
$\sqrt{225}$	$\sqrt{361}$	$\sqrt{289}$	$\sqrt{324}$

32) All kinds of Roots: Cubic, Quartic etc.

Common Radicals	Examples	What it means?
1. Square root (see the index is 2)	$\sqrt[2]{25} = 5$ $\longleftrightarrow$	when you multiply **5 to itself** it is 25 or $5 \times 5 = 25$ Or $5^2 = 25$
2. **Cubic root** (The index is 3).	$\sqrt[3]{8} = 2$ $\longleftrightarrow$	When you multiply 2 to **itself three times** the result is 8. or $2 \times 2 \times 2 = 8$ Or $2^3 = 8$
3. **Quartic Root**	$\sqrt[4]{81} = 3$ $\longleftrightarrow$	When you multiply 3 to itself **four times** the result is 81 **or** $3 \times 3 \times 3 \times 3 = 81$ Or $3^4 = 81$

Find the square, cubic or quartic roots as indicated

(1) $\sqrt[3]{1}$

2) $\sqrt[3]{8}$

(3) $\sqrt[3]{27}$

(4) $\sqrt[4]{16}$

(5) $\sqrt[4]{1}$

(6) $\sqrt[3]{125}$

(7) $\sqrt[4]{0}$

(8) $\sqrt{256}$

(9) $\sqrt[4]{256}$

(10) $\sqrt[3]{1000}$

(11) $\sqrt[3]{64}$

(12) $\sqrt{49}$

(13) $\sqrt{225}$

(14) $\sqrt[4]{81}$

(15) $\sqrt[4]{625}$

(16) $\sqrt[3]{216}$

33) Order of Operations: The Rules

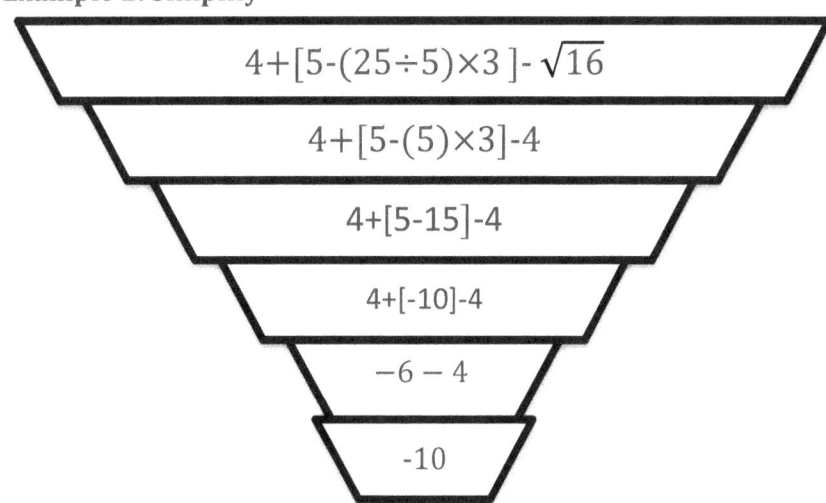

Without traffic law and order, there will be too many traffic crashes and chaos in the roads. Numbers also need some kind of order. **So, better learn the order of operations of numbers before causing number crash!**

To simplify mixed operations follow the order of operation.

Example 1: Simplify $(6 + 3 - 5)^2 \div (3 + 1) - (2 + 1) + 5 \times 6 - (2 \times 3) + \sqrt{16} =$

Step 1:	Do all operations within the **parenthesis**	$= (4)^2 \div (4) - (3) + 5 \times 6 - (6) + \sqrt{16}$
Step 2:	Work on **exponents**	$= 16 \div 4 - 3 + 5 \times 6 - 6 + \sqrt{16}$
Step 3:	Do the **roots**	$= 16 \div 4 - 3 + 5 \times 6 - 6 + 4$
Step 4:	Do the **multiplications and divisions** in order; from left to right	$= 4 - 3 + 30 - 6 + 4$
Step 5:	Finally, do all **additions and subtractions** in order: from left to right.	$= 1 + 30 - 6 + 4$
		$= 31 - 6 + 4$
		$= 29$

Example 2: Simplify

$$4+[5-(25\div5)\times3\,]-\sqrt{16}$$
$$4+[5-(5)\times3]-4$$
$$4+[5-15]-4$$
$$4+[-10]-4$$
$$-6-4$$
$$-10$$

34) Simplify Using the Order of Operations

1) $16 - 4 + 25 - 17$

2) $9 \div 3 \times 2$

3) $8 \div (4 \div 2^2) + 6$

4) $54 \div (32 - 31 - 3)$

5) $4 \times 2^3 - 2^2 (6 - 2) + 3^2$

6) $50 \div 5 \times 10 + 3$

7) $(7 - 5)^2 + 2$

8) $2^4 \div 6 - \sqrt{25}$

9) $16 \div 2(6 - 2)$

10) $12 \div 6 \div 2$

11) $2^4 + 7 - 9 \div 3$

12) $16 - 5(3)^2$

35) Advanced Order of Operations

1) $(5-3)^3 \times 2 - 5$

2) $3 \times [2 \times (6-3)^2] + 6(9 \div 3)^3$

3) $8 - 2[11 - (8 \div 2^2)] + 6$

4) $4)\ 50 \div 5 \times 10 + 3$

5) $10 \times 2^3 \div 4 - 21$

6) $6)\ 14 \div 7 \times (4 \div 2) \times 3$

7) $3 \times [2 \times (6-3)^2] + 6(9 \div 3)^3$

8) $7 + 3 \times 2^4 \div 5 + \sqrt{81}$

9) $392 \div (-4-3)^2 - 2(16 \div 8)^2$

10) $18 \div 3 \times 3 \div 2 \times 3\sqrt{16}$

36) Classification of Real Numbers

Real numbers are all the numbers. They are classified as **rational and irrational numbers**.

1) Rational Numbers include integers, fractions and decimals. **But, be careful: the decimals** should either repeat or terminate.

$$-5, \quad -1, \quad 0, \quad 10, \quad \frac{3}{4}, \quad \frac{1}{5}, \quad 0.25, \quad 0.161616$$

2) Integers are the numbers in the number line.

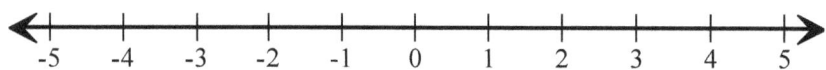

3) **Whole Numbers** are integers without the negative numbers.

$$0, 1, 2, 3, 4, 5, 6, 7, \ldots \ldots$$

4) Natural Numbers are whole numbers without the zero. They are called counting numbers.

$$1, 2, 3, 4, 5, 6, 7, \ldots \ldots$$

5) Irrational Numbers do not include integers or fractions. They include decimals that do not repeat or terminate (never ends).

$$\pi \text{ (read it as pi)} \qquad \sqrt{2} \qquad 1.456321321 \ldots$$

37) Real numbers Practice

Write either a rational (R) or Irrational (IR) next to each number. If it is rational write also what else the number is; Natural (N), Integer (I), or a Whole Number (W). Remember a number can be more than one thing. The first two were done for you!

1) 9 (R, I,W, N)

2) $\sqrt{5}$ (IR)

3) 0.0414141

4) -18

5) 0

6) 2π

7) $\sqrt{25}$

8) 4^2

9) -27

10) $\frac{4}{5}$

11) $\sqrt{3}$

12) 7.12512534

13) 0.1

14) -1248

15) $\sqrt{121}$

16) - $\sqrt{25}$

17) $\sqrt{7}$

18) 16 − 16

19) -27

20) $5\frac{2}{7}$

Commutative Property	$a + b = b + a$	$a \times b = b \times a$
	$5 + 2 = 2 + 5$	$3 \times 5 = 5 \times 3$
Associative Property	$a + (b + c) = c + (b + a)$	$a \times (b \times c) = b \times (a \times c)$
	$5 + (2 + 4) = 4 + (2 + 5)$	$3 \times (5 \times 4) = 5 \times (3 \times 4)$
Distributive Property	$a (b + c) = ab + ac$	
	$3(5 + 2) = 3 \times 5 + 3 \times 2$	

Name the property used:

1) $x + 3 = 3 + x$

2) $2 (3 + x) = 6 + 2x$

3) $y \times 3 = 3 \times y$

4) $4(x + y) = 4x + 4y$

5) $5y + x = x + 5y$

6) $(x + y) + z = (x + z) + y$

7) $8x \times 2 = 2 \times 8x$

8) $3b + a = a + 3b$

9) $ax + ay = a (x + y)$

10) $7x + 1 = 1 + 7x$

11) $4a(b \times c) = 4b(a \times c)$

12) $4a + (2b + c) = 2b + (4a + c)$

39) Algebraic Properties II

Additive Inverse Property	$a + (-a) = 0$	$7 + (-7) = 0$
Zero Property	$a \times 0 = 0$	$3 \times 0 = 0)$
Identity Property	$a + 0 = a$	$3 + 0 = 3$
Multiplicative Identity Property	$a \times 1 = a$	$7 \times 1 = 7$
Multiplicative Inverse Property	$a \times \left(\dfrac{1}{a}\right) = 1$	$7 \times \left(\dfrac{1}{7}\right) = 1$

Name the property used:

1) $x + 0 = x$

2) $15 \times 1 = 15$

3) $y + (-y) = 0$

4) $w \times \left(\dfrac{1}{w}\right) = 1$

5) $3b + 0 = 3b$

6) $(x + y) \times 0 =$

7) $8x + (-8x) = 0$

8) $1 = 9 \times \left(\dfrac{1}{9}\right)$

9) $1 \times (x + y) = x + 9$

10) $7x \times 0 = 0$

11) $2(x + y) = 2x + 2y$

12) $1 \times 15x = 15x$

40) Integers: Absolute Values

Integers: Include whole numbers and negative numbers.

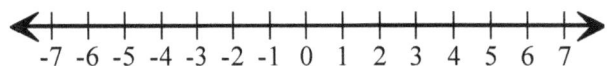

- **Gaining or more of something** is positive on number line.
- **Less of something is negative** in the number line.
- **The absolute value** of any number is always positive : $|-9| = 9$
- **For two negative numbers**, the one closer to the zero is greater: $-2 > -7$

Is the integer in each situation positive or negative?

1. The temperature of Minneapolis is 20 degrees below zero.

2. Amina has lost seventy five dollars.

3. Farah has gained a profit of $300.

4. Omar has lost 20 pounds before gaining 30 pounds.

5. A baby grows 9 inches taller.

6. Ali rides the hill for 200 feet.

7. Sofia has removed 10 books from the store.

Evaluate the expressions

8) $|-20|$

9) $|-12|$

10) $|-12| + |-13|$

11) $|-15| + |2|$

12) $-|-17|$

13) $|-12| + |10|$

14) $|-13| + |-6| - 5$

15) $28 - |-17|$

16) $|-12| + |10| - 10$

41) Ordering Integers: Which is bigger?

- **A positive number** is always greater than negative number: $4 > -10$
- **For two negative numbers**, the one closer to the zero is greater: $-2 > -5$
- **The absolute value** of any number is always positive: $|-7| = 7$

A) Use either a $<$ or $>$ to say the truth about a number:

1) -8 ◯ -3 5) -2 ◯ $-1/2$ 9) $-3/4$ ◯ $-1/2$

2) -3 ◯ -9 6) 0 ◯ -5 10) -4.5 ◯ -5.05

3) 0.5 ◯ -2 7) -17 ◯ -19 11) 2.5 ◯ -7.05

4) $|-7|$ ◯ 20 8) $|-6|$ ◯ -24 12) $|-12|$ ◯ 11

B) Order these numbers from smallest to largest

1) $-7, -5, 5, -9, 0, |-10|$ 2) $-4, -11, 15, -91, 0, |-15|$

3) $|-4|, -4, 0, -11, 13, |-14|$ 4) $|-14|, -14, 10, -12, 1, |-11|$

5) $-1, -5, |-7|, -4, -10, -11$ 6) $|-9|, -9, -10, -12, -21$

42) Add Integers

- **To add integers of the same signs:** add numbers and keep their sign.

    1) $(-2) + (-6) = -8$ **2)** $-5 + -6 = -11$ **3)** $(5) + (4) = 9$

- **To add integers of the different signs:** Just subtract and choose the sign of the larger number (in absolute value):

    1) $(2) + (-6) = -4$ **2)** $-8 + 3 = -5$ **3)** $-5 + 9 = 4$

1. $-6 + (-4)$	**11.** $9 + (-14)$	**21.** $-14 + 4 + (-5)$
2. $-9 + (-2)$	**12.** $-14 + 9$	**22.** $20 + (-14) + (-5)$
3. $8 + (-4)$	**13.** $-33 + (-4)$	**23.** $-5 + (-5) + (-7)$
4. $-1 + 5$	**14.** $-21 + (10)$	**24.** $-10 + 7 + (-5)$
5. $-13 + 4$	**15.** $-17 + (-4)$	**25.** $-20 + 6 + (-3)$
6. $-13 + (-4)$	**16.** $-16 + (5)$	**26.** $-10 + 10 + (-8)$
7. $-15 + 21$	**17.** $-20 + (-4)$	**27.** $26 + (-26) + (-8)$
8. $10 + (-14)$	**18.** $-13 + 22$	**28.** $0 + (-14) + (-5)$
9. $8 + (-4)$	**19.** $40 - 35$	**29.** $12 + (-10) + (-5)$
10. $14 - 12$	**20.** $-13 + (-4)$	**30.** $6 + (-4) + (-5)$

43) Subtract Integers

1) Connect **any two consecutive negatives as one positive, then simplify.** $$6 - (-4) = 6 + 4 = 10$$ $$12 - (-5) = 12 + 5 = 17$$	2) **Change** any consecutive positive and negatives into negative, then simplify. $$-9 - (+4) = -9 - 4 = -13$$ $$12 - (+4) = 12 - 4 = 8$$

1. $-7 - (-3)$

2. $-9 - (-2)$

3. $8 - (-4)$

4. $-1 - 5$

5. $-13 - 4$

6. $-13 + (-4)$

7. $-15 + 21$

8. $-13 - (-4$

9. $8 - (-4)$

10. $9 - 14$

11. $9 - (-14)$

12. $-14 - 9$

13. $-33 - (-4)$

14. $-21 - (10)$

15. $-17 - (-4)$

16. $-16 - (3)$

17. $-10 - (-4)$

18. $13 - (-2)$

19. $-40 - 35$

20. $-13 - (-4)$

21. $-10 - 4 - (-5)$

22. $-8 + (-14) - (-5)$

23. $-5 - (-5) + 7$

24. $11 + 4 - (-12)$

25. $-30 - 6 - (-36)$

26. $-10 - 10 - (-8)$

27. $-23 + (-26) - 8$

28. $8 - (-14) + (-5)$

29. $-12 - (-10) + (-5)$

30. $5 - (-6) - (-5)$

44) Add and Subtract Algebraic Expressions

✓ Combine like terms only: (**combine only x and x and not x and y**)

✓ *Remember also: it is usual to write (x) rather than (1x)*

Examples:
1) $4x - 5x = (4 - 5)x = -x$
2) $-5x + 5x = (5 - 5)x = 0$
3) $5x - 5y = 5x - 5y$
4) $8y - 6y = (8 - 6)y = 2y$

1. $-6x - 4x$

2. $-3y - 2y$

3. $8x - (-4y)$

4. $-x - 3x$

5. $-2a - 4b$

6. $3y - (-4y)$

7. $-13m - 4m$

8. $-10x - (-4x)$

9. $-8x - (9)$

10. $6x - (-4x)$

11. $19y - 14y$

12. $-w - 9w$

13. $-3t - 14t$

14. $-20x - x$

15. $-17t - 4t$

16. $-16x - (5)$

17. $-8x + 8x - 5x$

18. $6y - 20y + 5y$

19. $-12y - (-5y)$

20. $25k - (-4k) + 3y$

45) Evaluate Algebraic Expression I

Examples: **Evaluate Expressions if x= -2, y =3**		
1) $x - 5 =$ $-2 - 5 = $ **-7**	**2)** $-6 - x$ $= -6 - (-2)$ $= -6 + 2 = -4$	**3)** $2x - 2y + b =$ $2(-2) - 2(3) =$ $-4 - 6 = -10$

Directions: Evaluate Expressions if x= -2, y =3, a= -3 and b= 4

1. $6x - 4x$

2. $-3y - 2y$

3. $3x + 4y$

4. $-3x + 3x$

5. $-2a - 4b$

6. $3y - (-4y)$

7. $13x - 4x$

8. $-6x - (-4y)$

9. $-8x - (9)$

10. $6x - (-4x)$

11. $19y - 14y$

12. $-2a - 6b$

13. $-3x - 14y$

14. $14y - x$

15. $-7b - 4a$

16. $-16x - (5)$

17. $-4x + 6y - 2x$

18. $6y - 2y - 5x$

19. $-12y - (-5y)$

20. $5a - (-4b) + 3y$

46) Multiply Signed Integers

✓ **To multiply same sign** : Multiply the numbers and keep your answer positive:

      Examples: 1) $-6 \times -4 = 24$ 2) $(-3)(-5) = 15$ 3) $4 \times 5 = 20$

✓ **To multiply different signs:** Multiply numbers & keep your answer negative:

      Examples: 1) $-7 \times 3 = -21$ 2) $(5)(-2) = -10$ 3) $5(-7) = -35$

---

1. $-3 \times (-4)$

2. $(-9)(-2)$

3. $8 \times (-4)$

4. -1×5

5. -13×4

6. $-13 \times (-4)$

7. -15×-2

8. $10 \times (-2)$

9. $-1 \times (-4)$

10. 9×4

11. $9 \times (-1)$

12. -1×-9

13. $(-3)(-4)$

14. $-11 \times (3)$

15. -7×-4

16. $-16(5)$

17. $-8 \times 8 \times 5$

18. $-3 \times -4 \times -5$

19. $-1 \times -2 \times -5$

20. $-10 \times 2 \times -5$

47) Divide Signed Integers

To divide same sign integers : Divide the numbers and keep your **answer positive:**

Examples: 1) $-6 \div -2 = 3$ 2) $(-20) \div (-5) = 4$ 3) $14 \div 2 = 7$

To divide different sign integers: Divide numbers and keep the **answer negative:**

Examples: 1) $-16 \div 2 = -8$ 2) $(15) \div (-3) = -5$ 3) $35 \div (-7) = -35$

1. $-8 \div -4$

2. $-20 \div 4$

3. $8 \div (-4)$

4. $(-30) \div (-3)$

5. $(-9) \div (-3)$

6. $-12 \div (-4)$

7. $24 \div -4$

8. $-10 \div 10$

9. $-4 \div (-4)$

10. $-18 \div -2$

11. $9 \div (-1)$

12. $-7 \div -1$

13. $-5 \div 1$

14. $10 \div (-2)$

15. $-9 \div -1$

16. $-16 \div -4$

17. $-4 \div -2$

18. $-8 \div 8 \times 5$

19. $-20 \div -4 \times -5$

20. $7 - 12 \div (-4)$

48) Multiply & Divide Simple Algebraic Expressions

1. Positive × Positive = (+) 2. Negative × Negative = (+) 3. Negative × Positive = (−) 4. Positive × Negative = (−)	Example 1: $3 \times 2 = 6$ Example 2: $-3y \times -5 = 15y$ Example 3: $-3y \times 5 = -15y$ Example 4: $4y(-5x) = -20xy$
1. Positive ÷ Positive = (+) 2. Negative ÷ Negative = (+) 3. Negative ÷ Positive = (−) 4. Positive ÷ Negative = (−)	Example 1: $10 \div 2 = 5$ Example 2: $-10x \div -5 = 2x$ Example 3: $-15x \div 5 = -3x$ Example 4: $15x \div -5 = -3x$

Simplify Each Expression: **Remember** *the big dot* (•) *and the parenthesis ( ) mean times.*

1.　$-6 \cdot (-4)$

2.　$6x \div (-3)$

3.　$-3z \times -4y$

4.　$-4r(8s)$

5.　$-2a(-4b)$

6.　$9y \times (-3)$

7.　$-3(2) \times (-4xy)$

8.　$-4r \times (8s)$

9.　$-3(x)(-y)$

10.　$-2(-4c)$

11.　$-6 \cdot (-4x)$

12.　$-4x \cdot (-4)$

13.　$-4z(0)(xy)$

14.　$(-8)(-2)(-3x)$

15.　$(-5x)(-2)(3)$

16.　$-9 \cdot (-6y)$

17.　$-4(-4w)$

18.　$-6y(-4)$

19.　$-3w(-4)$

20.　$3(-2)(-2w)$

49) Evaluate Algebraic Expressions II

Example: Evaluate Expression if x= -2 and y =3

a) $-6x\,(5y) =$	b) $(2)(-4xy)=$
$-6(-2)(5\cdot 1) = 12(5)= 60$	$(2)\,(-4\cdot -2\cdot 3)= 2(8\cdot 3)= 2(24) =48$

Evaluate Each expression if   x= 3 ,   y=-1, a= 2 and b =-2

1) $-8\bullet(-4x)$

2) $9y\bullet(-3)$

3) $-6\bullet(-4x)$

4) $-9\bullet(-6y)$

5) $6x\div(-3)$

6) $-3(2)(-4xy)$

7) $-4x\bullet(-4)$

8) $-4b(-4a)$

9) $-3a\bullet-4y$

10) $-4a(8x)$

11) $-4xy$

12) $-6y\,(-4)$

13) $-4(8y)$

14) $-3(x)(-y)$

15) $(-8x)(-3x)$

16) $-3a(-4)$

17) $-2a(-4b)$

18) $-2a(-4b)$

19) $(-5x)(-2a)$

20) $(-2x)(-2a)$

50) Evaluate Algebraic Expressions III

To evaluate variables, replace the variables with the numbers and simplify:

Example: **Evaluate** $x(y+z)$ when $x = 5$, $y = 3$ and $Z = 1$

$$5(3+1) = 5 \times 4 = 20$$

Evaluate each expression if $x = 6$, $y = 3$, and $z = 2$.

1. $x + 2y + z$ 2. $3x - y$

3. $x + y - z$ 4. $3x - y + 3z$

5. $12z - x$ 6. $3(x + y + z)$

7. $xy \div z$ 8. $xyz - x$

Evaluate each expression if $a = 8$, $b = 4$, $c = 6$, and $d = 3$.

9. $a + b - c$ 10. $a + b - (c + d)$

11. $3a + 4d$ 12. $bc - d$

13. $(a + b) \div (c - d)$ 14. $c(4 + d)$

15. $ab - cd$ 16. $bc + a - d$

17. Suuban's age is two years more than twice of her brother's age.

   a. Write an expression for Suuban's Age.

   b. If Suuban's brother is currently 6 years, how old is Asha?

18. The total math grade for Mark, Abdi and Layla is 245.

   a. Using variables x, y and z, write an expression for the three students' score.

   b. If Mark and Abdi scored 170, write an expression for the total score.

51) Equations: Find the Solution

Example: Which number is the solution of the equation? $x + 15 = 21$; **4, 5, or 6?**

The Solution is **6**. Why? Just substitute **x** for **6** to have

$$6 + 15 = 21 \text{ OR } 21 = 21$$

Note: Other numbers such as **4** and **5** wouldn't work. Try them!

1) $x + 15 = 35$; $7, 8, 20$

2) $x - 15 = 5$; $10, 15, 20$

3) $x + 3 = 25$; $21, 22, 3$

4) $x - 7 = 14$; $28, 25, 21$

5) $4 + x = 11$; $7, 5, 6$

6) $7 + x = 13$; $6, 7, 8$

7) $18 - x = 3$; $14, 15, 16$

8) $21 - x = 7$; $14, 15, 18$

Find the Solution mentally:

9) $x + 20 = 21$

10) $x - 10 = 21$

11) $x + 17 = 22$

12) $x - 15 = 2$

13) $14 - x = 2$

14) $27 - x = 12$

15) $3 - x = 0$

16) $7 - x = 2$

52) Solve One Step Equations: Addition

Follow the examples to solve for x, y, or z:

1) $x + 15 = 35$
Isolate the x by subtracting 15 from both sides: $$x + 15 - (15) = 35 - (15)$$ $$x = 20$$

2) $x + 3 = 12$
Isolate the x by subtracting 3 from both sides: $$x + 3 - (3) = 12 - (3)$$ $$x = 9$$

1) $x + 4 = 10$ **2)** $x + 9 = 12$ **3)** $x + 2 = -1$ **4)** $x + 7 = -4$

5) $x + 3 = 10$ **6)** $y + 8 = 0$ **7)** $x + 16 = 10$ **8)** $x + 9 = 1$

9) $x + 14 = 14$ **10)** $x + 3 = 11$ **11)** $x + 8 = -1$ **12)** $y + 9 = 11$

13) $x + 7 = 11$ **14)** $z + 11 = 4$ **15)** $x + 8 = -12$ **16)** $x + 9 = -5$

17) $y + 4 - 3 = 11$ **18)** $x + 4 + 2 = 10$ **19)** $z + 1 + 2 = 12$ **20)** $x + 9 = 0$

53) Solve One Step Equations: Subtraction

Follow the examples to solve for x, y, or z:

1) $x - 9 = 10$
Isolate the x by adding 9 to both sides: $x - 9 + (9) = 10 + (9)$ $x = 19$

2) $x - 5 = -5$
Isolate the x by adding 5 to both sides: $x - 5 + (5) = -5 + (5)$ $x = 0$

1) $x - 4 = 10$ 2) $x - 2 = -10$ 3) $x - 3 = -1$ 4) $x - 7 = -4$

5) $x - 3 = 12$ 6) $y - 9 = 0$ 7) $x - 6 = -10$ 8) $x - 9 = -9$

9) $x - 12 = 10$ 10) $x - 3 = 10$ 11) $x - 8 = -1$ 12) $y - 9 = 11$

13) $x - 7 = -11$ 14) $z + 1 = -4$ 15) $x - 8 = -2$ 16) $x - 2 = -5$

17) $y - 3 - 6 = 11$ 18) $x + 1 - 2 = -3$ 19) $z - 1 - 2 = -12$ 20) $x - 8 = 0$

54) One Step Equations: Addition & Subtraction

Follow the examples to solve for x, y, or z:

1) $-9 + x = -1$
Isolate the x by adding 9 to both sides: $$x - 9 + (9) = -1 + (9)$$ $$x = 8$$

2) $-5 = 8 + x$
*Isolate the x by subtracting **8** from both sides:* $$-5 - (8) = 8 - 8 + x$$ $$-13 = x$$

1) $x + 5 = 1$ **2)** $x - 3 = -6$ **3)** $3 + x = -7$ **4)** $4 - 7 = x$

5) $-9 + y = -2$ **6)** $y + 9 = 6$ **7)** $y - 6 = -10$ **8)** $2 + x = -14$

9) $-x - 8 = 4$ **10)** $-13 = x + 4$ **11)** $x - 8 = -1$ **12)** $y - 9 = 11$

13) $x - 7 = -11$ **14)** $z + 1 = -4$ **15)** $x + 8 = -12$ **16)** $x - 3 = -8$

17) $12 + x - 6 = 10$ **18)** $x - 2 + 5 = -10$ **19)** $z + 8 - 10 = -1$ **20)** $x - 15 = 0$

55) Solve One Step Equations: Multiplication

Example 1: $8x = 16$	$\dfrac{8x}{8} = \dfrac{16}{8}$	$x = 2$	
Example 2: $5x = -20;$	$\dfrac{5x}{5} = \dfrac{-20}{5}$	$x = -4$	
Example 3: $-3x = -1;$	$\dfrac{-3x}{-3} = \dfrac{-1}{-3}$	$x = \dfrac{1}{3}$	
Example 4: $-x = 10;$	$\dfrac{-x}{-1} = \dfrac{10}{-1}$	$x = -10$	

1) $2x = 10$ **2)** $3x = 9$ **3)** $2x = -10$ **4)** $4x = -4$

5) $-2x = -12$ **6)** $-6y = -18$ **7)** $-3y = -24$ **8)** $-7x = -21$

9) $-4x = -14$ **10)** $-3x = -1$ **11)** $-5x = 10$ **12)** $2y = -14$

13) $8x = -16$ **14)** $-z = 4$ **15)** $-x = -1$ **16)** $-x = 5$

17) $-4y = 36$ **18)** $-x = 0$ **19)** $2y = -1$ **20)** $-32y = -8$

56) Solve One Step Equations: Division

Example 1:	$\dfrac{x}{2} = 12$	$x = 2 \times 12$	$x = 24$
Example 2:	$\dfrac{x}{5} = -3$	$x = 5 \times -3$	$x = -15$
Example 3:	$\dfrac{x}{-4} = -9$	$x = -4 \times -9$	$x = 36$
Example 4:	$\dfrac{-x}{2} = 10$	$-x = 2 \times 10$	$x = -20$

1) $\dfrac{x}{5} = 2$

2) $\dfrac{x}{4} = -2$

3) $\dfrac{x}{8} = -12$

4) $\dfrac{x}{-5} = 7$

5) $\dfrac{x}{-4} = -9$

6) $\dfrac{x}{-6} = -2$

7) $\dfrac{x}{-3} = -12$

8) $\dfrac{x}{-7} = -7$

9) $\dfrac{-x}{7} = -2$

10) $\dfrac{x}{-9} = -7$

11) $\dfrac{-x}{8} = -2$

12) $\dfrac{x}{-12} = 2$

13) $\dfrac{x}{-16} = 2$

14) $\dfrac{x}{-8} = 0$

15) $\dfrac{x}{15} = -8$

16) $\dfrac{x}{9} = -8$

17) $\dfrac{x}{5} = 2$

18) $\dfrac{-x}{3} = 11$

19) $\dfrac{x}{-25} = 0$

20) $\dfrac{x}{-2} = 10$

57) One Step Multiplication/Division Equations

Follow the examples to solve for x, y, or z:

1) $15 = -5x$	**2)** $\dfrac{x}{4} = 8$
Isolate the x by dividing -5 to both sides $\dfrac{15}{-5} = \dfrac{-5x}{-5}$ *or* $-3 = x$	*Isolate the x by multiplying 4 to both sides:* $\dfrac{4 \times x}{4} = 4 \times 8$ *or* $x = 32$

1) $4x = 16$

2) $3x = -1$

3) $\dfrac{x}{-2} = -7$

4) $-5x = 15$

5) $\dfrac{x}{9} = -1$

6) $x + 3 = -27$

7) $-x = -1$

8) $-5 = \dfrac{x}{5}$

9) $-11 = \dfrac{x}{-5}$

10) $20 = y - 8$

11) $\dfrac{x}{-7} = 0$

12) $-y = 11$

13) $-\dfrac{x}{8} = 2$

14) $20x = -5$

15) $2x = -10$

16) $\dfrac{x}{-6} = -9$

17) $-x = -51$

18) $27 = -3y$

19) $30x = -5$

20) $-7x = -5$

58) Solve One Step Equations: Fractions

Follow the examples to solve for x, y, or z:

$$1)\ x + \frac{1}{7} = \frac{5}{7}$$	$$2)\ x - \frac{1}{2} = \frac{3}{8}$$
Isolate the x by subtracting $\frac{1}{7}$ from both sides and then simplify the fraction as needed: $$x + \frac{1}{7} - \frac{1}{7} = \frac{5}{7} - \frac{1}{7} \quad or\ x = \frac{5-1}{7} = \frac{4}{7}$$	*Isolate the x by adding $-\frac{1}{2}$ to both sides and then simplify the fractions:* $$x - \frac{1}{2} + \frac{1}{2} = \frac{3}{8} + \frac{1}{2} \quad or\ x = \frac{3}{8} + \frac{1}{2}$$ $$x = \frac{3}{8} + \frac{1 \times 4}{2 \times 4} \quad or\ x = \frac{7}{8}$$

$$1)\ x + \frac{2}{5} = \frac{3}{5} \qquad\qquad 2)\ x + \frac{2}{9} = \frac{5}{9} \qquad\qquad 3)\ x + \frac{1}{7} = \frac{5}{7}$$

$$4)\ x - \frac{1}{5} = \frac{3}{10} \qquad\qquad 5)\ x - \frac{5}{6} = \frac{1}{2} \qquad\qquad 6)\ x + \frac{1}{2} = \frac{3}{5}$$

$$7)\ x + \frac{3}{8} = \frac{1}{5} \qquad\qquad 8)\ x + \frac{3}{7} = \frac{11}{21} \qquad\qquad 9)\ 4 = x - \frac{3}{7}$$

$$10)\ \frac{1}{3} = x - \frac{3}{5} \qquad\qquad 11)\ x + \frac{1}{4} = 0 \qquad\qquad 12)\ x + \frac{1}{3} = \frac{3}{7}$$

59) Solve One Step Equations: Decimals

Follow the examples to solve for x, y, or z:

1) $15 = 0.5x$	**2) $x - 1.7 = 8.3$**
Isolate the x by dividing 0.5 to both sides $$\frac{15}{0.5} = \frac{0.5x}{0.5} \quad or \quad 30 = x$$	Isolate the x by adding 1.7 from both sides: $$x - 1.7 + 1.7 = 8.3 + 1.7 \quad or \quad x = 10$$

1) $2x = 1.6$ **2)** $0.33x = 0.66$ **3)** $\dfrac{x}{-2} = 0.7$ **4)** $x - 1.5 = 3$

5) $\dfrac{x}{0.2} = 2.5$ **6)** $x + 3.7 = -2.8$ **7)** $0.1\,x = -1$ **8)** $-0.2 = \dfrac{x}{5}$

9) $-10 = \dfrac{x}{1.2}$ **10)** $10.5 = y - 2.8$ **11)** $\dfrac{x}{-7} = 0.1$ **12)** $-y = 1.9$

13) $\dfrac{x}{0.4} = -0.25$ **14)** $-2.5x = -12.5$ **15)** $0.02x = -10$ **16)** $\dfrac{x}{6} = 0.5$

17) $0.3x = -5.1$ **18)** $2.7 = 3y$ **19)** $x + 12.5 = -7.2$ **20)** $-7x = 3.5$

60) Solve One Step Equations: Mixed Review

1) $-3x = 21$

2) $x + 3 = -10$

3) $\dfrac{x}{-3} = -6$

4) $-5x = 5$

5) $\dfrac{x}{8} = -12$

6) $3x = -30$

7) $2 = -10x$

8) $\dfrac{x}{5} = 2$

9) $-13 = \dfrac{x}{5}$

10) $2 = y + 3$

11) $\dfrac{x}{-2} = 0$

12) $-y = -17$

13) $\dfrac{x}{4} = -6$

14) $20 + x = 4$

15) $x + \dfrac{1}{6} = \dfrac{1}{3}$

16) $-2 = \dfrac{x}{-7}$

17) $21 + x = -1$

18) $-4 = -4 + y$

19) $x - \dfrac{7}{9} = \dfrac{1}{3}$

20) $-7x = 35$

21) $5.1 + x = -1.3$

22) $= \dfrac{x}{0.8} = -1.2$

23) $2.8 + x = 11$

24) $-0.5x = 5$

61) Math Language: Translating Math Expressions

Addition Words: add, plus, sum, increased, all, total, altogether….	**Subtraction Words:** subtract, minus, difference, decreased by, reduced, less…
Multiplication: product, times, twice, multiplied by etc.	**Division:** divide, average, shared evenly or equally, split equally, quotient…

Translate the following math phrases into numerical expressions:

Example: eleven more than twenty $11 + 20$

1. the difference between thirty and fifteen

2. the sum of five, four and three

3. the product of twenty and two

4. fifteen less than 70

5. the quotient of 15 and 7

6. seventeen increased by five

7. eight reduced by seven

8. the quotient of twenty and 5 decreased by four

Translate into verbal expressions:

Example: $11 \times 5 + 20$; the product of eleven and five increased by twenty

9. $5 - 3$

10. $7 + 3$

11. $5/9$

12. $5 \times 3 \div 7$

13. $33 + 2 + 7$

14. $5 - 3$

62) Math Language: Variables

Variables: In math letters such as **x, y, z** etc. are used to represent the unknown variables.

    Example: a number increased by 5 is represented as **x + 5**.

Translate each phrase into an algebraic expression. Use letter "x" as the unknown variable. Use also y for the second variable if there is any.

1. eleven more than a number

2. the difference between a number and fifteen

3. twenty pounds more than his weight

4. the product of five and a number

5. fifteen less than the product of five and number

7. the quotient of 7 and a number

8. seventeen increased by ten times a number

9. nine more than the product of three and two

10. the quotient of twenty and 5 decreased by four.

11. twice of her age increased by twice of her age

12. eight times the product of length and width

63) Equations: Word Problems

Example: the sum of 5 and a number is 25. What is the number?
1) Build the equation: $5 + x = 25$
2) Solve: since $5 + 20 = 25$; x must equal 20

Write the equation and then solve.

1) The sum of a number and 3 is ten

2) Four times a number increased by 2 is 22.

3) The combined age of Ali and Asha is 27. Ali is 14. How old is Asha?

4) A number increased by 35 is 62.

5) A number decreased by 10 is 42.

6) Five times a number is 45.

7) The quotient of 12 and a number 4.

8) Since last year, a baby grows 13 inches to become 30 inches. How tall was the baby?

9) The product of a number and 7 is 63.

10) Omar was 150 pounds a week ago, then he lost 3 pounds before gaining more weight to become 180 pounds. How much he gained?

64) Two Step Equations: Variables on One Side

Follow the examples to solve for x, y, or z:

1) $2x + 15 = -5$
Move the **15** to the other side by subtracting **15** from both sides then solve: $2x + 15 - 15 = -5 - 15$ **or** $2x = -20$ *Therefore*: $\dfrac{2x}{2} = \dfrac{-20}{2}$ *or* $x = -10$

2) $\dfrac{x}{4} + 8 = 3$
Move the **8** to the other side by subtracting **8** from both sides then solve $\dfrac{x}{4} + 8 - 8 = 3 - 8$ $\dfrac{x}{4} = -5 \, or \, x = -20$

1) $2x + 7 = 21$

2) $3x + 2 = -10$

3) $\dfrac{x}{-3} + 5 = -2$

4) $5x - 5 = 5$

5) $\dfrac{x}{5} + 8 = -1$

6) $3x - 3 = -30$

7) $2 = -10x + 12$

8) $\dfrac{x}{6} - 13 = 2$

9) $-10 = \dfrac{x}{5} + 4$

10) $21 = 8y + 3$

11) $\dfrac{x}{-2} + 12 = 0$

12) $6y + 3 = 9$

13) $15x + 40 = -5$

14) $20 + 8x = 4$

15) $4x + 5 = -55$

16) $2 = \dfrac{x}{-7} + 1$

Follow the examples to simplify the algebraic expressions:

1) $2(3-5x) =$	2) $-3(2x-5y) =$
Multiply (distribute) the 2 to both terms: $2 \times 3 + 2(-5x) = 6 - 10x$	Multiply (distribute) the -3 to both terms : $-3 \times 2x + (-3 \times -5y) = -6x + 15y$

Use distributive property to write the equivalent algebraic expression:
Remember *the* the parenthesis ( ) mean times.

1. $-6(x-4)$

2. $6(-3+x)$

3. $-3(z-4y)$

4. $-4(r+8s)$

5. $-2(2a-4b)$

6. $9(-3+2x)$

7. $-3(2x-4y)$

8. $-4(8-3y)$

9. $-3(x-y)$

10. $-5(k-4m)$

11. $3(5-4x)$

12. $-4(2x-8)$

13. $-4(x+y)$

14. $(-2-3x)5$

15. $-5(-2y-3)$

16. $-9(-6y-5)$

17. $(7-4w)(-4)$

18. $-2(-6y-4x)$

19. $(x-4)(-3)$

20. $3(-2-2w)$

66) Combining Like Terms

Follow the examples to combine like terms

1) $2x + 15 + 5x = 7x + 15$	2) $15x - 20x + 5y + 3 - 7$ $= -5x + 5y - 4$
2x & **5x** are like terms. So add their coefficients $(2+5)x = 7x$ **15 is by itself.**	15 & -20 are like terms: $(15 - 20)x = 5x$ 3 and -7 are like terms = -4 **5y** is by itself.

1) $2x + 7x - 21$ 2) $8x - 2y - 4x$ 3) $y - 5y$ 4) $3x - 5x + 2y$

5) $2(-3 + 2x) - 4x$ 6) $-3x - 3x + 1$ 7) $-2(-10x - 3) + x$ 8) $5x - 8x - 2$

9) $-3x + 4x - y$ 10) $-2y + 8y + 2x$ 11) $-2(y + 2 - y)$ 12) $6y - 3 + 9$

13) $2(x - 3) - 4$ 14) $20x + 8x - 4x$ 15) $-4y - 2y + 10y$ 16) $3x - 1 - 4x$

17) $-2(a - 3b) - 4a$ 18) $16x^2 + 8x - 4$ 19) $-4y + 2y + 10y^2$ 20) $-x - 1 - 7x$

67) Add/ Subtract Algebraic Fractions

Follow the example to simplify the algebraic Fractions:

Example	1) $\dfrac{x}{3} - \dfrac{x}{5}$
1. Find the LCM. 2. Make the denominators equal to the LCM by multiplying right numbers. 3. Add/ Subtract and Simplify	The LCM is **15** $\dfrac{x}{3} - \dfrac{x}{5} = \dfrac{x \times ⑤}{2 \times ⑤} + \dfrac{x \times ③}{5 \times ③}$ $= \dfrac{5x - 3x}{15} = \dfrac{2x}{15}$

1. $\dfrac{5x}{2} - \dfrac{x}{2}$

2. $\dfrac{x}{3} - \dfrac{x}{4}$

3. $\dfrac{2x}{3} + \dfrac{x}{9}$

4. $\dfrac{3x}{5} - \dfrac{2x}{5}$

5. $\dfrac{4y}{7} - \dfrac{2y}{5}$

6. $\dfrac{3w}{4} - \dfrac{w}{5}$

7. $\dfrac{9t}{7} + \dfrac{2t}{14}$

8. $\dfrac{3k}{4} - \dfrac{2k}{3}$

9. $\dfrac{t}{2} + \dfrac{2t}{5}$

10. $\dfrac{2x}{7} - \dfrac{2x}{5}$

11. $\dfrac{m}{3} - \dfrac{7m}{4}$

12. $\dfrac{2x}{4} - \dfrac{5x}{16}$

13. $\dfrac{t}{3} + \dfrac{t}{3}$

14. $\dfrac{5x}{9} - \dfrac{x}{4}$

15. $\dfrac{n}{7} - \dfrac{n}{7}$

16. $\dfrac{4x}{3} - \dfrac{5x}{4}$

68) Solve Equations: Variables on Both Sides I

Follow the example to solve equations:

Example: Solve for x:	$9x = 3x + 12$	$2 + x = 3x + 12$
Step1: Get the variable on one side. Subtract the smaller from both sides. Step 2: Simplify	$9x - (3x) = 3x - (3x) + 12$ $6x = 12$ $x = 2$	$2 + x - (x) = 3x - (x) + 12$ $2 = 2x + 12$ $2 - (12) = 2x + 12 - (12)$ $-10 = 2x$ $-5 = x$

1) $7x = 2x + 10$ 2) $3x = 14 - 4x$ 3) $y - 5y = 12$ 4) $3x = 5x + 16$

5) $2(-4 + x) = 4x$ 6) $-x + 4 = 3x + 12$ 7) $-2(5x - 3) = 2x$ 8) $5x = 8x - 3$

9) $3x = 4x + 1$ 10) $-2y = 8y + 2$ 11) $-2(y + 2) = 2y$ 12) $6y = 3 + 9y$

13) $2(x - 3) = 4x$ 14) $20x = 8x - 24$ 15) $4y - 2y = 10y + 8$ 16) $3x - 1 = 4x$

17) $-2(a - 3) = 4a$ 18) $16 + 8x = 4x$ 19) $4y + 2 = 10y + 4$ 20) $-x - 1 = 7x$

69) Solve Multistep Equations: A Review

Example 1 $2y + 4 = 7 + 5y$	Example 2: $3(y - 3) = -27$
$2y - (2y) + 4 = 7 + 5y - (2y)$ $4 = 7 + 3y$ $4 - \mathbf{7} = 7 - \mathbf{7} + 3y$ $-3 = 3y$ $-1 = y$	$3y - 9 = 27$ $\frac{3}{3}y = \frac{-18}{3}$ $3y - 9 + 9 = -27$ $+ 9$ $\boxed{y = -6}$ $3y = -18$

1) $\dfrac{3x}{4} = 4$

2) $2x + 4 + 3x = 4x + 1$

3) $\dfrac{5x}{4} = \dfrac{x}{2} + 5$

4) $\dfrac{x}{-3} = -8\,(-2)$

5) $n + n + 7 = 27$

6) $0.2\,y + 0.8y = 1.6$

7) $2x + 6 = -7x + 15$

8) $\dfrac{5}{4} + y = 1 + \dfrac{5}{4}$

9) $\dfrac{m}{2} = \dfrac{1}{4} + \dfrac{2}{8}$

10) $\dfrac{x}{4} = \dfrac{1}{4} + \dfrac{5}{2}$

11) $23x - 12 = 3x$

12) $7t - 40 = 3t + 20$

13) $\dfrac{4t}{0.04} = \dfrac{1}{0.8}$

14) $3(7 - t) = 3(2t + 1)$

15) $3 + 3(n - 1) = -21$

70) Solve and Graph Inequality

1) Solve Inequality just like you do in the equality but use the inequality sign:

Less <,   **Less or equal ≤,**   **Greater >, Greater or Equal ≥**

2) See these examples:

1) $x + 3 \geq 1$ $x + 3 + (-3) \geq 1 + (-3)$
   $Or \ x \geq -2$

   See the interval is closed (dark)

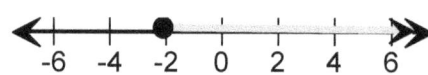

2) $-4 < x \leq 2$ See how one side is open and the other side is closed

3) $-2x \leq -8$ $\dfrac{-2x}{-2} \geq \dfrac{-8}{-2} \ \ or \ \ x \geq 4$

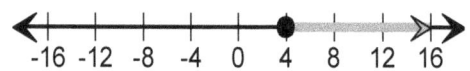

Note how the inequality **changes direction** when multiplied or divided by negatives.

Solve (if needed), then graph the inequality

1) $x \geq 1$

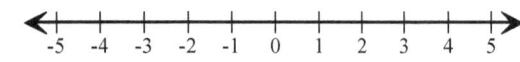

2) $y - 5y \leq 12$

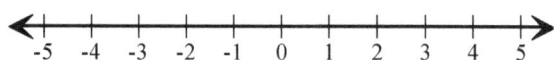

3) $-3 < x \leq 1$

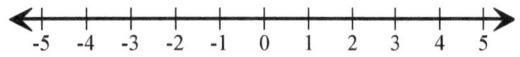

4) $5x < 8x - 3$

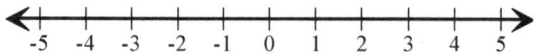

5) $3x + 6 > 2x + 1$

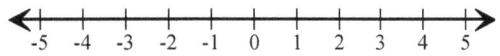

6) $-2 \leq x \leq 1$

71) Write the corresponding Inequality

1) x > 2

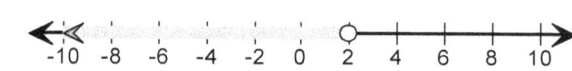

2)

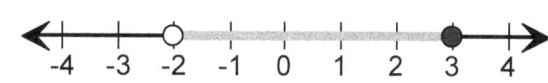

3)

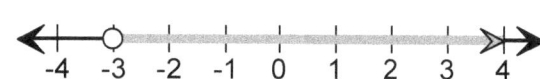

4)

5)

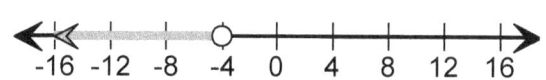

6)

7)

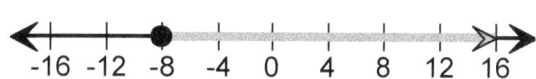

8)

9)

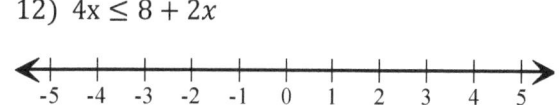

10)

Solve and draw the inequality

11) $x + 5 \geq 1$

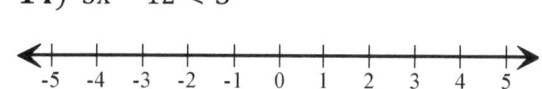

12) $4x \leq 8 + 2x$

13) $-2w - 4 \leq 4$

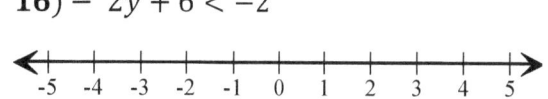

14) $3x - 12 < 3$

15) $-3x - 6x > -18$

16) $-2y + 6 < -2$

72) Exponents: The Basics

---

1. An exponent represents repeated multiplications:

- ✓ Instead of writing: $2 \times 2 \times 2 \times 2$, Simply write: 2^4
- ✓ Instead of writing: $y.y.y.y.y$, Simply write y^5

2. The base and the power:

- ✓ The **2** and **y** are called bases
- ✓ The **4 & 5** are the exponents

3. Read it :

- ✓ 5^3 as, five to the third power
- ✓ y^4 as, y to the fourth power

4. Zero power:

- ✓ *Any number that has **zero** as an exponent is 1*
- ✓ *Therefore,* $(-127)^0 = 1$; $y^0 = 1$; *any number$^0 = 1$*

5. To add or subtract variables:

- ✓ Add only same variables with the same power. $3x^3 + 2x^3 = 5x^3$
- ✓ Never add a variable and a number: $2x^2 + 5 = 2x^2 + 5$

6. Never add two powers:

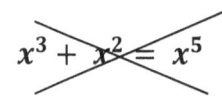

$x^3 + x^2 = x^5$

Exercise: Simplify

1) $9x^2 + 4x^2 =$

2) $5y^2 - 3y^2 + y =$

3) $2x^3 + x^0 + 5 =$

4) $3^3 + 2x^3 + 5x^3 =$

5) $10y^2 - 3y^2 + 2x^2 =$

6) $7^3 - 5x^2 + x =$

73) Add or Subtract Exponents

Example 1: $3^4 + x^2$ $= 3 \times 3 \times 3 \times 3 + x^2$ $= 81 + x^2$	**Example 2:** $5^2 + 7^2$ $= 5 \times 5 + 7 \times 7$ $= 25 + 49$ $= 74$	**Example 3:** $3x^2 + 6x^2 =$ $(3 + 6)x^2 = 9x^2$

(1) $10^0 =$

(2) $3^0 + 1^0 + x^0 =$

(3) $5^3 - 1^0 =$

(4) $3^4 - 3^0 =$

(5) $5^2 - 6 =$

(6) $9x^2 + 7x^2 =$

(7) $5y^7 + 4y^7 =$

(8) $0^5 + 2^2 =$

(9) $3^5 + 2^2 =$

(10) $5y + 3y^2 - 2y =$

(11) $0^5 + 10y^4 - 7y^4 =$

(12) $4^4 + 2^2 + 17^0 =$

(13) $10y^4 - 17y^4 =$

(14) $b^5 + 10y^4 - b^5 =$

(15) $4y^3 + 2y^2 + 17y^2 =$

74) Multiply and Divide Exponents

Simplify the Exponents. Follow the examples:

$2^3 \times 2^2$ ⇒ $2^{3+2} = 2^5$	$x^5 (x^3)$ ⇒ $x^{5+3} = x^8$
$\dfrac{3^7}{3^2} =$ ⇒ $3^{7-2} = 3^5$	$\dfrac{y^8}{y^3} =$ ⇒ $y^{8-3} = y^5$

(1) $3^4 \times 3^4 =$

(2) $x^4 \times x^8 =$

(3) $\dfrac{7^3}{7} =$

(4) $10^0 \times 3^4 =$

(5) $12^0 \times 12^3 =$

(6) $\dfrac{3^7}{3^2} =$

(7) $y^0 \times y^4 =$

(8) $\dfrac{m^7}{m^3} =$

(9) $5^2 - \dfrac{10^2}{10} =$

(10) $5 \times 3^2 - 27^0 =$

(11) $\dfrac{3^7}{3^2} - 3^5 =$

(12) $\dfrac{x^7}{4^3} \times \dfrac{x^7}{4^2} =$

75) Power of Power (Exponent of an Exponent)

Simplify the Exponents. Follow the examples:

$(x^3)^2 = \Rightarrow x^{3 \times 2} = x^6$	$x^5(x^3)^2 \Rightarrow x^5(x^6) = x^{11}$
$(y^4)^2 = \Rightarrow y^{4 \times 2} = y^8$	$\dfrac{y^{12}}{(y^5)^2} = \Rightarrow y^{12-10} = y^2$

(1) $(x^3)^3 =$

(2) $(x^5)^2 =$

(3) $(y^3)^2 =$

(4) $(x^5)^0 =$

(5) $(y^0)^2 =$

(6) $(2^3)^2 =$

(7) $\dfrac{y^6}{(y^2)^2} =$

(8) $3^2(2^2)^2 =$

(9) $x^5(x^5)^2 =$

(10) $2^0(2^4)^2 =$

(11) $(3^2)^2 =$

(12) $\dfrac{y^{12}}{(y^5)^0} =$

76) Negative Exponents

Simplify the Exponents and change all **negative exponents** into positive. Follow the examples:

$$y^{-4} = \frac{1}{y^4}$$

$$2^{-3} = \frac{1}{2^3} = \frac{1}{8}$$

$$y^{-4} \times y^7 = y^{-4+7} = y^3$$

$$\frac{x^{-5}}{x^2} = \frac{1}{x^2} \times \frac{1}{x^5} = \frac{1}{x^7}$$

(1) $x^{-3} =$

(2) $(x^{-5})^2 =$

(3) $(y^{-3})^2 =$

(4) $\frac{1}{y^{-9}} =$

(5) $(y^{-4})^0 =$

(6) $(2^{-3})^{-2} =$

(7) $\frac{x^{-5}}{x^{-2}} =$

(8) $\frac{x^{-5}}{x^{-2}} =$

(9) $\frac{x^{-7}}{x^{-2}} =$

(10) $\frac{x^9}{x^{-5}} =$

(11) $\frac{x^{-5}}{x^{-5}} =$

(12) $\frac{x^{-5}}{x^{-6}} =$

77) Evaluate Exponents

Evaluate the following exponents. Follow the examples:

$\Rightarrow$ $x^3 \times 3^2$ if $x = 2$	$\Rightarrow$ $\dfrac{y^7}{y^4}$ when $y = 3$
$= (2)^3 \times 3^2$ $= 2 \times 2 \times 2 + 3 \times 3$ $= 8 + 9$ $= 17$	$y^{7-4} = y^3$ $= 3^3$ $= 27$

(1) $x^4 = $ _____
   if $x = 3$

(2) $\dfrac{m}{7} = $ _____
   if $m = 21$

(3) $m^5 \times 2 = $ _____
   if m=2

(4) $\dfrac{x^7}{3^2} = $ _____
   when x=1

(5) $y^0 \times y^4 = $ _____
   when y =4

(6) $5^2 \times \dfrac{10^2}{10} = $ _____

(7) $n + n^2 - 8 = $ _____
   When $n = 4$

(8) $\dfrac{x^6}{x^5} \times \dfrac{x^3}{x^2} = $ _____
   When x = 4

(9) $\dfrac{m^5}{m^2} - 81^0 = $ _____
   If m=9

10) $\dfrac{n^4}{n^2} + n^2 = $ _____
   If n=5

78) Scientific Notation-1

Express each number in standard form. See examples below:

	Examples
Since the exponent (4) is positive move the decimal point 4 times to the right. Add zero if you need it.	1) $3.52 \times 10^4 = 35200$ 2) $2.643 \times 10^4 = 26430$
Since the exponents is negative move the decimal point 3 times to the left and 2 times to the left for the second example.	3) $1.47 \times 10^{-3} = 0.00147$ 4) $8.31 \times 10^{-2} = 0.0831$

1. 3.52×10^4

2. 2.8×10^2

3. 5.61×10^{-4}

4. 9.63×10^{-3}

5. 8.02×10^3

6. 7.12×10^4

7. 4.6×10^{-1}

8. 3×10^{-5}

9. 6.5×10^5

10) 5.22×10^6

10. 8.213×10^6

11. 1.61×10^{-5}

13) 3.5×10^{-6}

14) 9.13×10^0

15) 4.21×10^0

16) 3.0×10^1

79) Scientific Notation-2

Express each number in scientific notation. See examples below:

	Examples
First, put a decimal next to the first number. Since the decimal has moved three places to the left make the exponent 3. **Remember** the decimal for **2643 is hidden at the end. So,** it is indeed **2643.0**	$3521.8 = 3.5218 \times 10^3$ $2643 = 2.643 \times 10^3$
First, put a decimal next to the first number. Since the decimal has moved **four places to the right** make the exponent -4.	$0.000147 = 1.47 \times 10^{-4} =$ $0.000831 = 8.31 \times 10^{-4} =$

1. 0.00045

2. 2.8×10^2

3. 41,000

4. 310

5. 0.0021

6. 1000

7. 610,000

8. 45

9. 0.0032

10. 0.000021

11. 210

12. 0.00031

13. 0.0000013

14. 410

15. 81000

16. 0.56

To evaluate variables, such as x, y, z etc, replace the variables with the numbers and simplify.

Example 1	Example 2
$\sqrt{x^4}$ if x = 3	$2\sqrt{m} + 3\sqrt{m} =$ if m=9
$\sqrt{3^4} = 3^2 = 9$	$2\sqrt{9} + 3\sqrt{9} = 2 \times 3 + 3 \times 3$ $= 6 + 9 = 15$

(1) $\sqrt{x^2} =$ if x = 3	(2) $\sqrt{3x^3} =$ x= 3	(3) $\sqrt[3]{x^3} =$ if x = 7
(4) $\sqrt{36x^2} =$ if x = −1	(5) $\sqrt{m^4}$ when m=2	(6) $\sqrt{y^{10}}$ when y= 1
(7) $y^0 \cdot \sqrt{y^4} =$ when y =3	(8) $\sqrt{t} + 12\,t$ if t =16	(9) $10\sqrt{y} + 12\sqrt{y}$ When y =4
(10) $5\sqrt{x} - \sqrt{x} + 4\sqrt{x}$ When x = 4	(11) $8 \times \sqrt{y}$ When y =25	(12) $10 \times \sqrt{y} - 3 \times \sqrt{y}$ When y =9
(13) $\frac{4y \times \sqrt{y}}{9 \times \sqrt{y}} - 39^0 =$ if y=9	14) $2(\sqrt{t} - 4) =$ If t= 64	15) $10\sqrt{y} - 4\sqrt{y} =$ If y=5

1. A ratio compares two quantities of the same kind.

Example 1, if there are 3 boys and 5 girls in a classroom we say:

"The ratio of boys to girls is 3 to 5. Or the ratio of girls to boys is 5 to 3".

2. **Three ways to write ratios of boys to girls in the classroom:**

   ✓ We can us use the word **to**: The ratio of boys **to** girls is **3 to 5.**
   ✓ We can use the **colon**: The ratio of boys to girls is **3 : 5**
   ✓ Or we can use the **fraction**: The ratio of boys to girls is $\frac{3}{5}$

3) **If you write your answer as a fraction, write the first number at the top (numerator).**

Example 2: (a).There are 7 boys and 3 girls in a classroom.
The ratio of girls to boys is $\frac{3}{7}$.

Example 3: There are 10 blue crayons and 6 red crayons.
The ratio of the red crayons to blue crayons is $\frac{6}{10} = \frac{3}{5}$

(**Note we have simplified the fractions.**).

4) **To see if two ratios are equal**: Check if their cross products are equal

Example 4: Write >, or < or = to compare these ratios:

a)

$$\frac{1}{2} \bigcirc \frac{3}{6}$$

$$\frac{1}{2} \diagtimes \frac{3}{6}$$

6×1	=	2×3
or $6 = 6$		

b)

$$\frac{6}{17} \bigcirc \frac{7}{19}$$

$$\frac{6}{17} \diagtimes \frac{7}{19}$$

6×19	7×17
or $114 < 119$	

82) Write the Ratios & Simplify Your Answer

1. Jamal has a bag with 4 books, 8 pencils and 5 pens. What is the ratio of (a) books to pens? (b) Pens to pencils?

2. One in four students loves to play chess. In a class of 20 students, how many students love to play chess?

3. The foreign born persons in 2006 - 2010 were 7% in Minnesota and 13% for the US. What was the ratio of foreign born in Minnesota to that of the USA in 2006-2010?

(**Source:** US Census Bureau, Quick Facts, Minnesota, 2010)

Q 4 – 7: There are132 rooms in the White House including 35 bathrooms. There are 412 doors, 147 windows, 28 fireplaces, 7 staircases, and 3 elevators in the White House. (**Source:** www.whitehouse.gov)

4) What is the ratio of rooms to bathrooms?

5) What is the ratio of doors to windows?

6) What is the ratio of fireplaces to staircases?

7) What is the ratio of staircases and elevators to fireplaces?

8) Write equal (=) or greater (>) or less (<) for the following ratios:

(a) $\frac{3}{4}$( ) $\frac{3}{4}$ b) $\frac{15}{7}$( ) $\frac{20}{8}$ c) $2\frac{1}{5}$( ) $\frac{22}{10}$ d) 14( ) $\frac{28}{5}$

83) Proportions

1) ⇨ A proportion represents two ratios (fractions) that are equal:

$$\frac{6}{7} = \frac{12}{14}$$

2) **We can write proportions in two ways:**

1. As fractions: $\frac{a}{b} = \frac{c}{d}$

2. By using a colon:
$$a : b = c : d$$

3) **How to solve a proportion:** Just cross multiply the diagonal numbers and divide the results to the third number:

Example: Solve the following proportions:

a) $\frac{2}{5} = \frac{6}{x}$	b) $\frac{y}{5} = \frac{9}{7}$	c) $\frac{9}{2} = \frac{p}{6}$
$\frac{2}{5} \diagup\hspace{-1.2em}\diagdown \frac{6}{x}$	$\frac{y}{5} \diagup\hspace{-1.2em}\diagdown \frac{9}{7}$	$\frac{9}{2} \diagup\hspace{-1.2em}\diagdown \frac{p}{6}$
$2x = 6 \cdot 5$	$7y = 5 \cdot 9$	$2p = 9 \cdot 6$
$x = 30 \div 2$	$y = 45 \div 7$	$p = 54 \div 2$
$x = 15$	$y = \dfrac{45}{7}$	$p = 27$

84) Solve the following Proportions:

1) $\dfrac{x}{4} = \dfrac{7}{2}$

2) $\dfrac{x}{8} = \dfrac{7}{2}$

3) $\dfrac{x}{1.2} = \dfrac{0.5}{2}$

4) $\dfrac{2}{x} = \dfrac{7}{13}$

5) $\dfrac{20}{x} = \dfrac{6}{5}$

6) $\dfrac{3}{y} = \dfrac{8.2}{1.2}$

7) $\dfrac{9}{4} = \dfrac{y}{2}$

8) $\dfrac{5}{4} = \dfrac{y}{2}$

9) $\dfrac{9}{13} = \dfrac{y}{2}$

10) $\dfrac{3}{4} = \dfrac{12}{y}$

11) $\dfrac{19}{5} = \dfrac{21}{y}$

12) $\dfrac{1}{4} = \dfrac{7}{y}$

13) $\dfrac{t}{0.04} = \dfrac{1}{0.8}$

14) $\dfrac{9}{5} = \dfrac{9}{n}$

15) $\dfrac{n}{0.4} = \dfrac{7}{4}$

16) $8 = \dfrac{12}{r}$

17) $\dfrac{1.8}{5} = \dfrac{r}{7}$

18) $\dfrac{11}{4} = \dfrac{8}{r}$

85) How to Solve Proportion Word Problems:

Example 1: Suban runs each morning 4 miles in 15 minutes. How many minutes will it take her to run 18 miles?

Step1: Use variable, like x, d etc. for the missing number and set up the proportion (4 miles in 15 Minutes means divide).	$\dfrac{4 \text{ miles}}{15 \text{ minutes}} = \dfrac{18 \text{ miles}}{x \text{ minutes}}$
Step 2: **Cross Multiply:**	$15 \times 18 = 4\,x$
Step3: **Solve the proportion** (Isolate variable X)	$\dfrac{15 \times 18}{4} = \dfrac{4x}{4}$ $x = 67.5 \; minutes$

We can exchange denominators and numerators as far as the corresponding units are in the same side). Try this and call me if you do not get the same answer as before!)

$$\frac{15 \text{minutes}}{4 \text{ miles}} = \frac{x \text{ minutes}}{18 \text{ miles}}$$

Example 2:

A scale drawing 2 inches represent 30 miles. What distance does a line segment of 7 inches represent?

Step1: Set up the proportion (Let's call d the distance)	$\dfrac{2 \text{ inches}}{30 \text{ miles}} = \dfrac{7 \text{ inches}}{d}$
Step 2: Cross Multiply:	$2d = 7 \times 30$
Step3: Solve the proportion (Isolate the d)	$\dfrac{2d}{2} = \dfrac{210}{2}$ $d = 105 \; miles$

86) Solve the following Proportion Problems:

1) There was just 2 computers shared by every 5 students. If there were 32 students in class, how many computers they share?

2) Seven students out of every 10 Pre-university students who took the math test passed the test. This month 75 will take the test. How many are expected to pass the test.

3) It takes 2 cubs of sugar to prepare 50 cakes. How many cakes that can be prepared from five and half cubs of Sugar?

4) On a typical school day 5 students in every 80 students claim to have lost their pencils. How many students that will claim to have lost their pencils in a school of 500 students?

5) Gloria runs 6 miles in 30 minutes. At that rate, how far could she run in 120 minutes?

6) Assume test grades are proportional to study time. Asma has studied 4 hours to score of 80. How many hours should Asma study math per week if she wants to get 98?

7) Waris is preparing cake. She knows that every six cups of flour needs one cup of sugar. If she wants to use 36 cups of flour, how many cups of sugar she should used?

8) Jack drove his Honda for 200 miles and used 5 gallons of fuel. How long he traveled if he used 18 gallons?

Do you see from both the table and the graph how many minutes exercised when 12 calories were burned? In math, it is also common to use the **ordered pairs** as below:

(1, 2), (2, 4), (3, 6), (4, 8), (5, 10), (5, 10)

x Minutes in Stair Climber	y Calories Burned
1	2
2	4
3	6
4	8
5	10
6	12

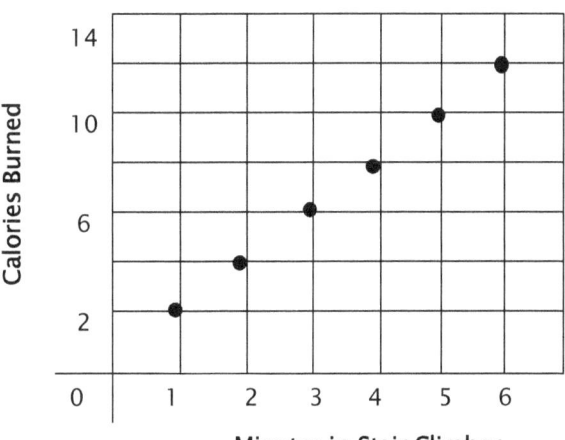

✓ The first number of the ordered pair is the **x-coordinate**. The second number of the ordered pair is called the **y-coordinate.**

✓ **The Coordinate System**: sometimes also called the Cartesian System is the system that puts ordered pairs in the graph.

✓ *The x-axis and y-axis* divide the system into four regions called Quadrants;

✓ **The Origin**: is where the x and y-axis meet and is zero: **(0,0)**

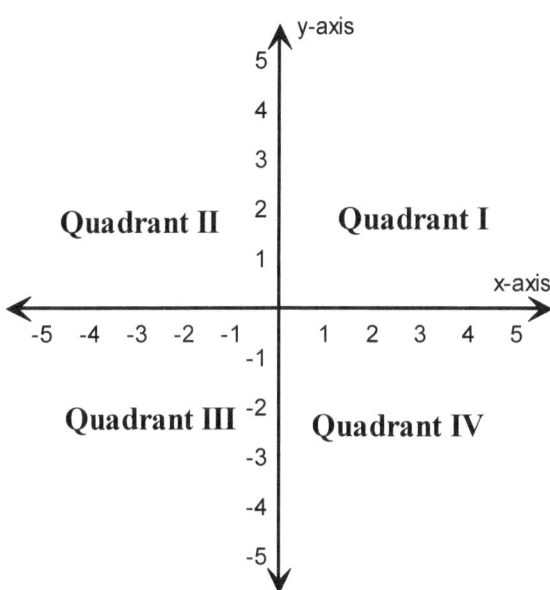

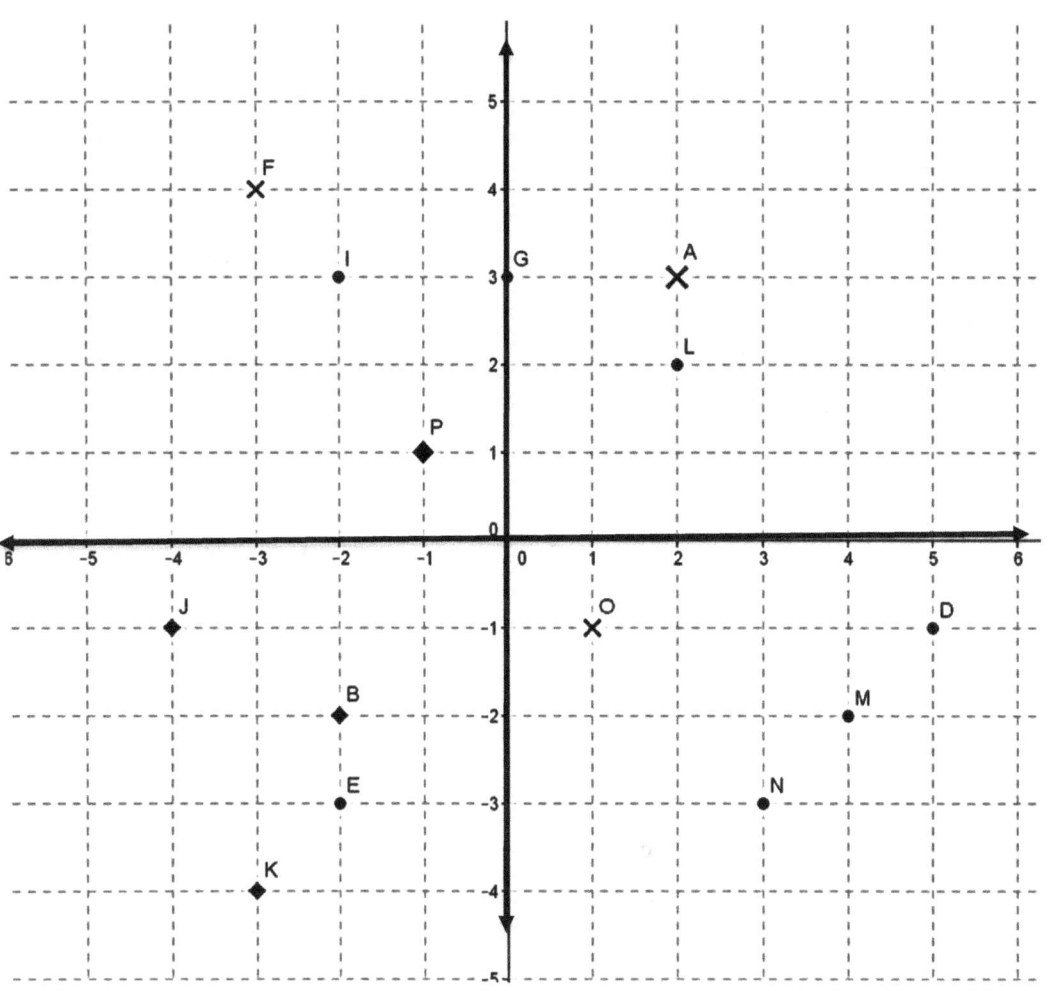

Write the ordered pairs shown in the graph:

A (2, 3)   B (,)   D (,)   E (,)   F (,)

G (,)   I (,)   J (,)   K (,)   L (,)

M (,)   N (,)   O (,)   P (,)

89) Graph ordered pairs in the Coordinate System

1. A (3, 4)

2. B (3, -2)

3. C (3, 1)

4. D (2, 4)

5. E (-3, 4)

6. F (-2, 3)

7. G (-5, 1)

8. H (0, 4)

9. I (3, 0)

10. J (-3, -4)

11. K (-1, -2)

12. L (-4, 4)

13. M (-6, 4)

14. N(-5, 3)

15. O (-5, 4)

16. P (3, 4)

17. Q (3, 4)

18. R (3, 4)

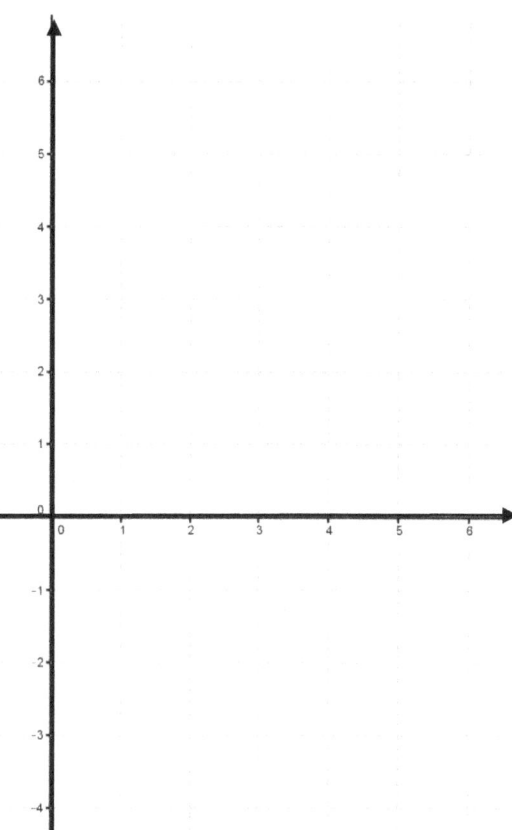

To find the ordered pairs of equation, give any value you like to x, and then solve the y. (Sorry,

Example : 2x +y = 8		
If we choose x=0	it means 2 (0) + y = 8	y = 8
If we choose x=1	it means 2(1) + y = 8	y = 6

you can't choose both x and y at the same time).

1) 2x +y =8		
x	y	(x, y)
0	8	(0, 8)
1	6	(1, 6)
2	4	(2, 4)
3	2	(3, 2)

2) x +y =5		
x	y	(x, y)

3) x -2y =6		
x	y	(x, y)
-2	-4	(-2,-4)
0	?	(0,-3)
1	-2.5	(1,-2.5)
2	?	(2,-2)

4) 2x +3y =6		
x	y	(x, y)

5) 2x -y =4		
x	y	(x, y)
-2		(-2,-8)
0		(0,-4)
1		(1,-2)
2		(2,0)

6) x +2y =-6		
x	y	(x, y)
-1		
0		
1		
2		

7) 2x +3y =6		
x	y	(x, y)
-2		
-1		
0		
1		

8) -2x -4y =8		
x	y	(x,y)
-2		
0		
1		
2		

9) 2x +y =-2		
x	y	(x, y)
0	8	(0, 8)
1	6	(1, 6)
2	4	(2, 4)
3	2	(3, 2)

10) x +y =0		
x	y	(x, y)
-1		
-1		
0		
3		

Prepare 2-3 ordered pairs, 2) **Plot** the point 3) **Graph** the line.

Example 1: **Graph y = 2x+3**

Let's find few ordered pairs and plot on the graph:

1) y =2x +3		
x	y	(x, y)
0	3	(0, 3)
-1	1	(1, 1)
1	5	(1, 5)

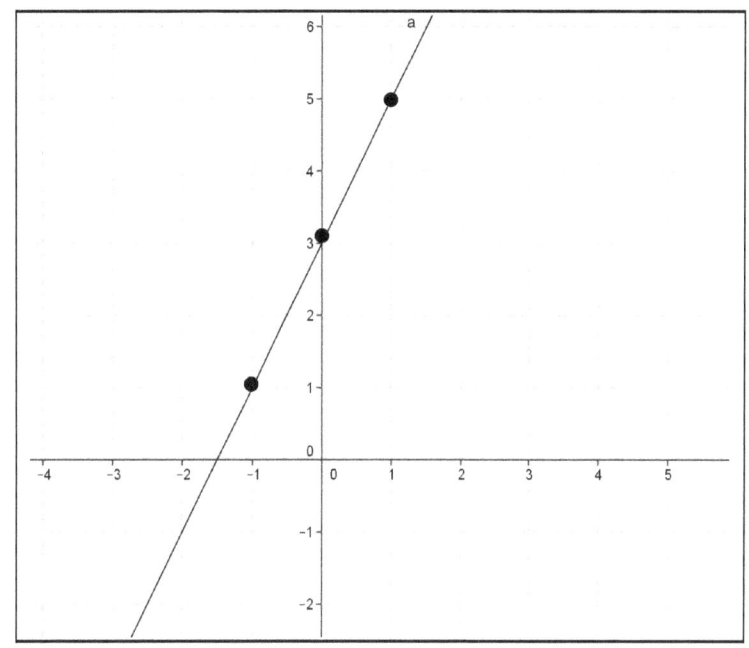

Practice: You plot this on the same graph:

y = x+1

Example 2: **Graph y = -4x+2**

y =-2x +2

x	y	(x, y)
0	2	(0, 3)
-1	4	(-1, 4)
1	0	(1, 0)

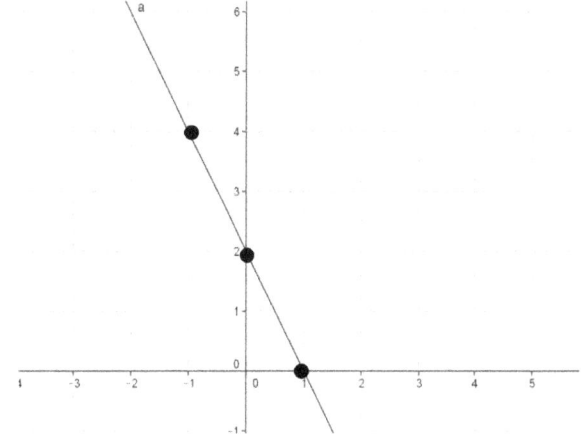

92) Graph These Linear Equations

1. **Prepare** 3 ordered pairs, 2) **Plot** the points 3) **Graph** the line:

1) $y = 2x$

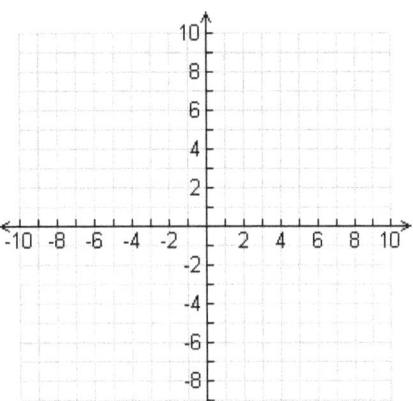

2) $y = x - 2$

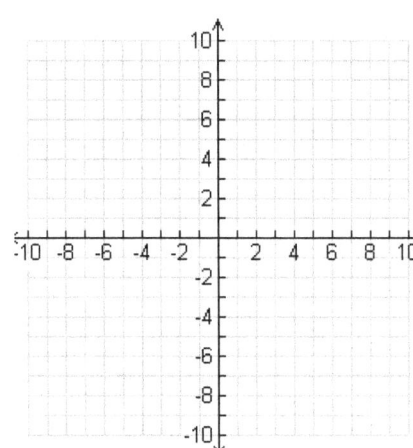

3) $y = -4x - 2$

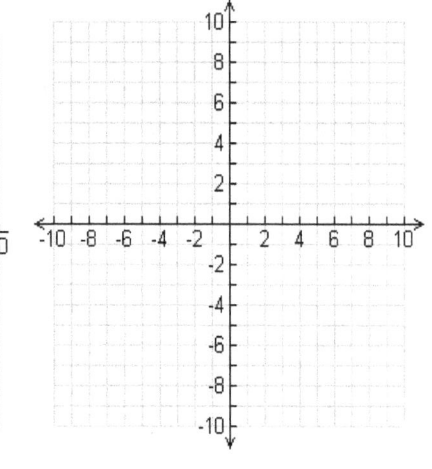

4) $y = x + 2$

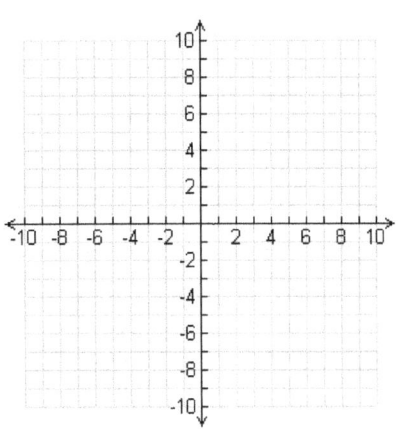

5) $y = 3x + 1$

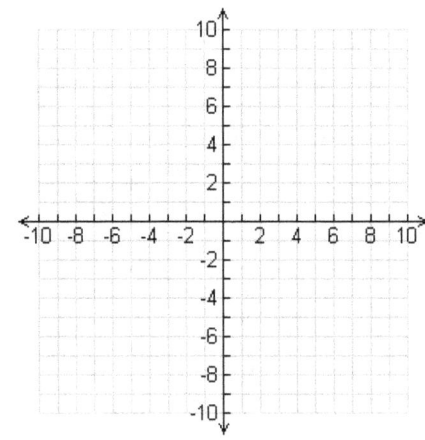

6) $2x + 36 = 6$

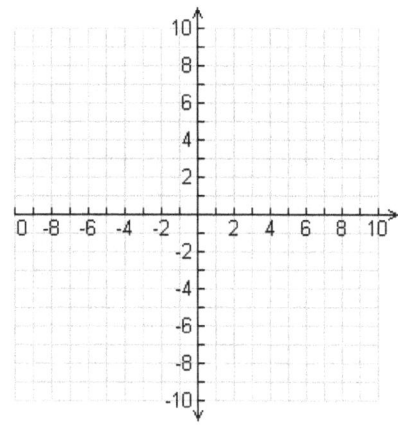

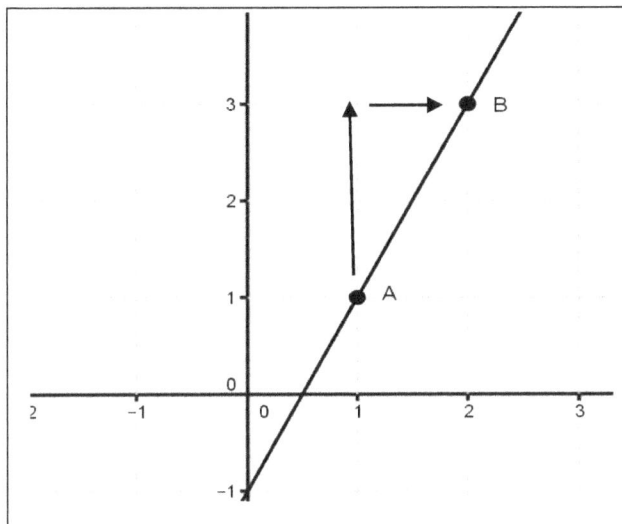

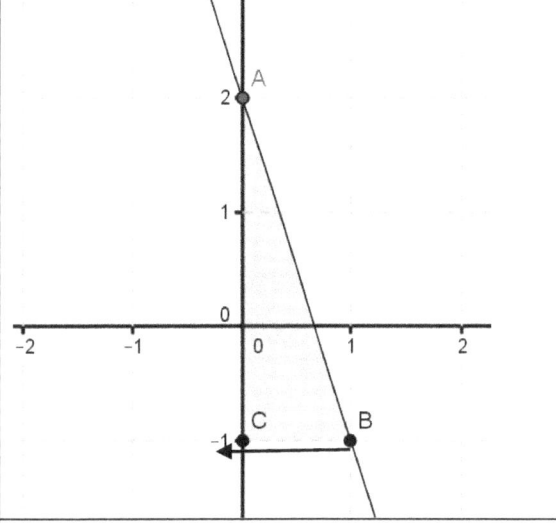

To get from point A to B rise up (at y axis) by 2 points and run (on x axis) by 1:

$$Slope = \frac{Rise}{Run} = \frac{2}{1} = 2$$

Rise from point C and A. = 3
Run from B to C= -1 (It is negative to move backwards)

$$Slope = \frac{Rise}{Run} = \frac{3}{-1} = -3$$

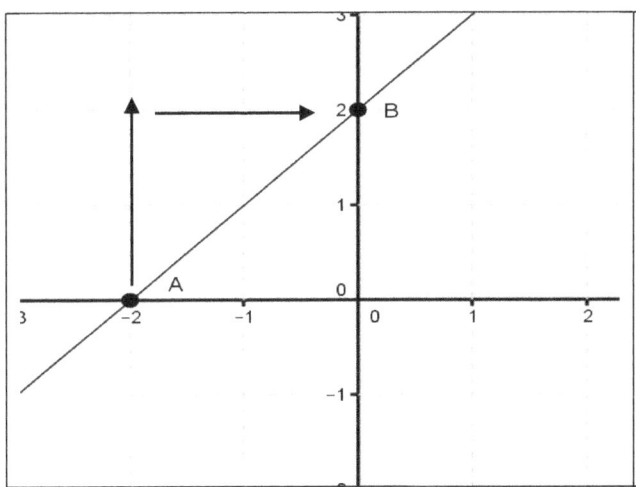

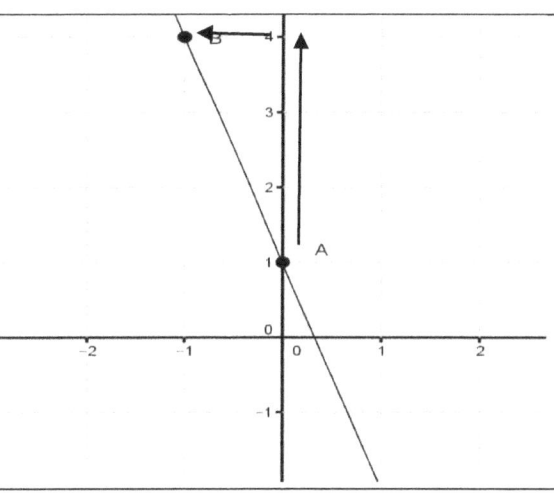

Rise from point A to B. = 2
Run for A to B= 2

$$Slope = \frac{Rise}{Run} = \frac{2}{2} = 1$$

$$\frac{Rise}{Run} = \frac{3}{-1} = -3$$

1)

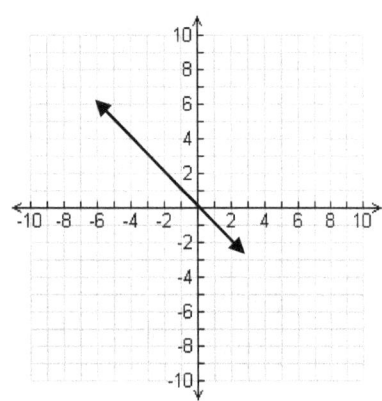

2)

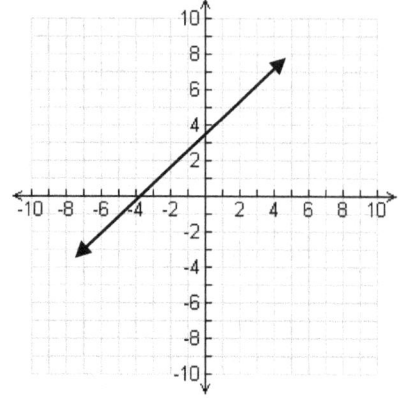

3)

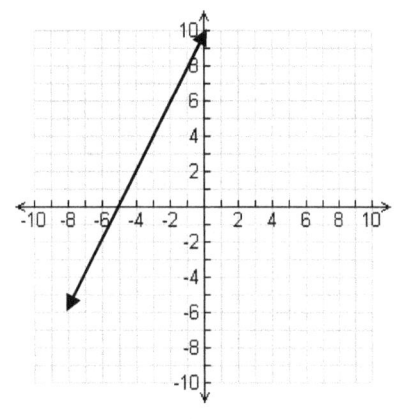

4)

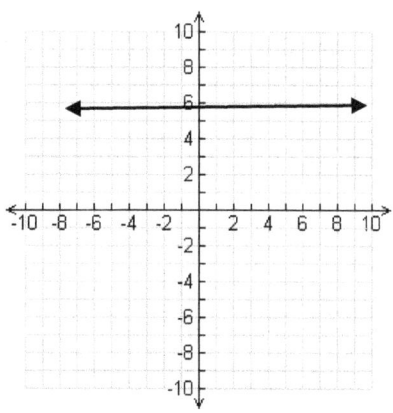

5)

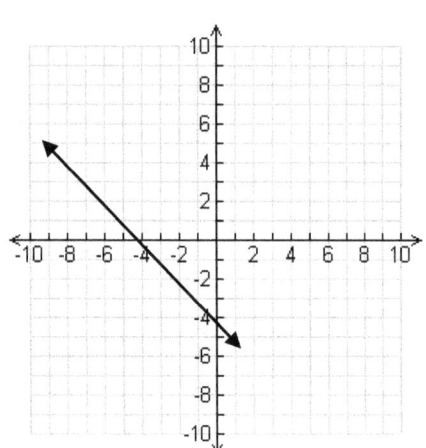

6)

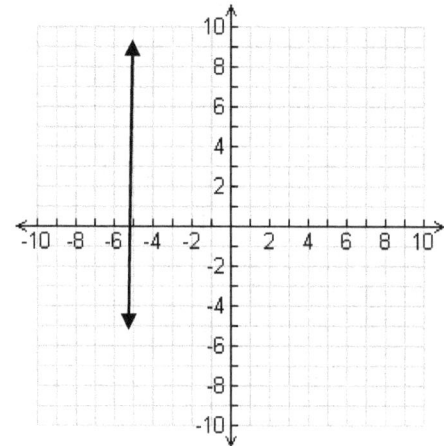

95) Find the Slope from Ordered Pairs

Find the slope that passes through each pair of ordered points: See examples.

	Example1 : (O,3),(4,1)	Example 2: (2,-3), (-4,6)
1) Label it:	$x_{1=0}, x_{2=4}$ $y_{1=3}, y_{2=1}$	$x_{1=2}, x_{2=-4}$ $y_{1=-3}, y_{2=6}$
2) Put it in the Formula: $m = \dfrac{y_2 - y_1}{x_2 - x_1}$	$m = \dfrac{1-3}{4-0} = \dfrac{-2}{4} = -\dfrac{1}{2}$	$m = \dfrac{6-(-3)}{-4-2} = \dfrac{9}{-6} = -\dfrac{3}{2}$

Remember: m stands for **slope.** Also, watch out for the signs!

1. (2, 3), (4,1) _____   (2,-0), (-1, 2) _____

2. (-4, 3), (5,1) _____   (2,7) , (7, 6) _____

3. (1, 3), (2,4) _____   (7,-3), (6, 6) _____

4. (5, 5), (6, 7) _____   (7,-3), (0, 6) _____

5. (3, 3), (-4, 1) _____   (2,-3), (-4, 6) _____

6. (2,-3), (4, 7) _____   (5,-3), (3, 6) _____

7. (3,-3), (-4, 6) _____   (2,-3), (1, 2) _____

8. (7,-3), (-4, 8) _____   (4,-3), (3, 8) _____

9. (1, 4), (4, 7) _____   (2,-3), (-4, 9) _____

10. (0,-3), (-4, 5) _____   (2, 9), (-5, 6) _____

96) Write the Equation of the Line: Use Slope/Y-intercept:

A) Use the given slope and the y-intercept to **write the equation of the line** in Slope Intercept form $(y = mx + b)$. See examples:

Example1 :	Example 2 :
Slope =3 y-intercept =1	Slope =-1 y-intercept =0
$y = mx + b$ $y = 3x + 1$	$y = mx + b$ $y = -x + 0$ Or simply: $y = -x$

Remember: "m" is the slope and "b" stands for the y-intercept.

1) Slope =2, y-intercept = 1

2) Slope =-2, y-intercept = 2

3) Slope =-4, y-Intercept = -2

4) Slope =-1, y-intercept = 7

5) Slope = 5, y-Intercept = 0

6) Slope =-6, y-intercept = 11

7) Slope = 0, y-Intercept = -5

8) Slope $=\frac{2}{3}$, y-intercept = -38

9) Slope = -1, y-Intercept = -3

10) Slope =-10 , y-Intercept = -27

B) Given the equation of the line, find the slope and the Y-intercept. See Example:

11) $y = 3x + 1$	$m = 3;\ b = 1$	12) $y = 7$	
13) $y = -5x + 7$		14) $y = 8x$	
15) $y = -6x - 4$		16) $y = -3x - 1$	
17) $y = x$		18) $y = 2x + 3$	
19) $y = -9x + 8$		20) $y = -4x - 5$	

Example: Graph the line with the given equation;
$$y = 2x - 1$$
First: see the slope =2 and the y-intercept =-1
So, mark the y-intercept (or -1).

Next: go up 2 points and over 1 point to the right.
1) Connect the point (dots)!

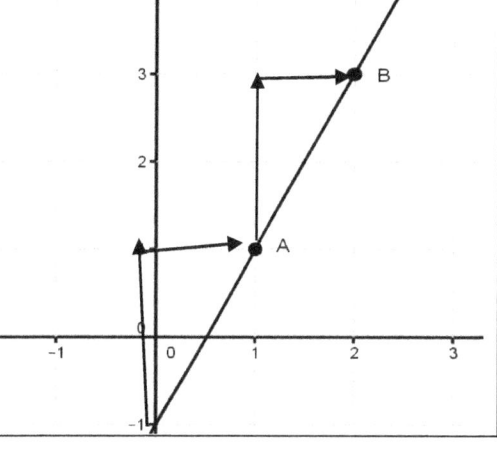

$y = 4x - 2$

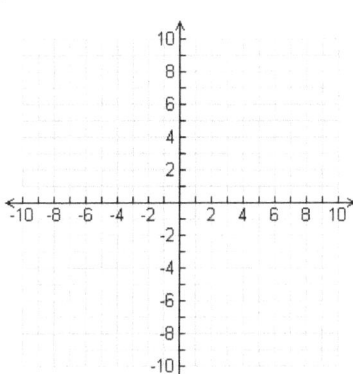

$y = x + 1$

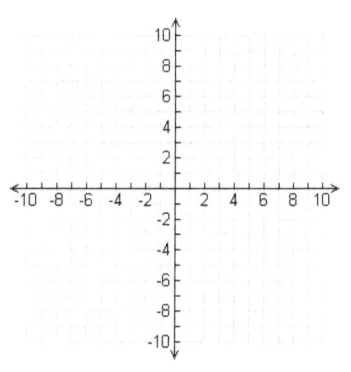

$y = 2x + 2$

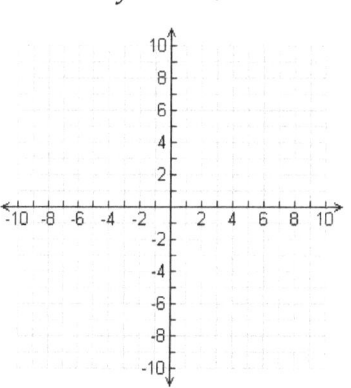

$y = -2x - 1$

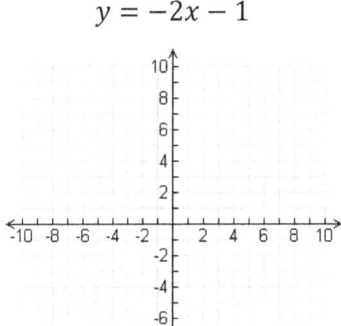

$y = 3x$

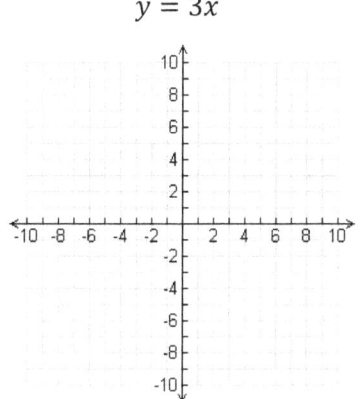

$y = 6x - 6$

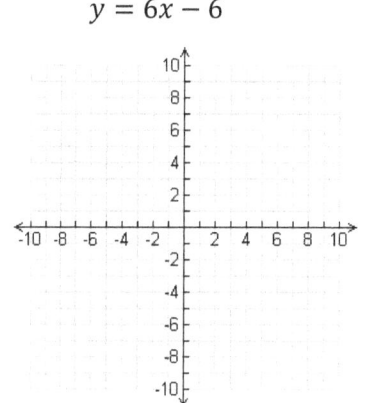

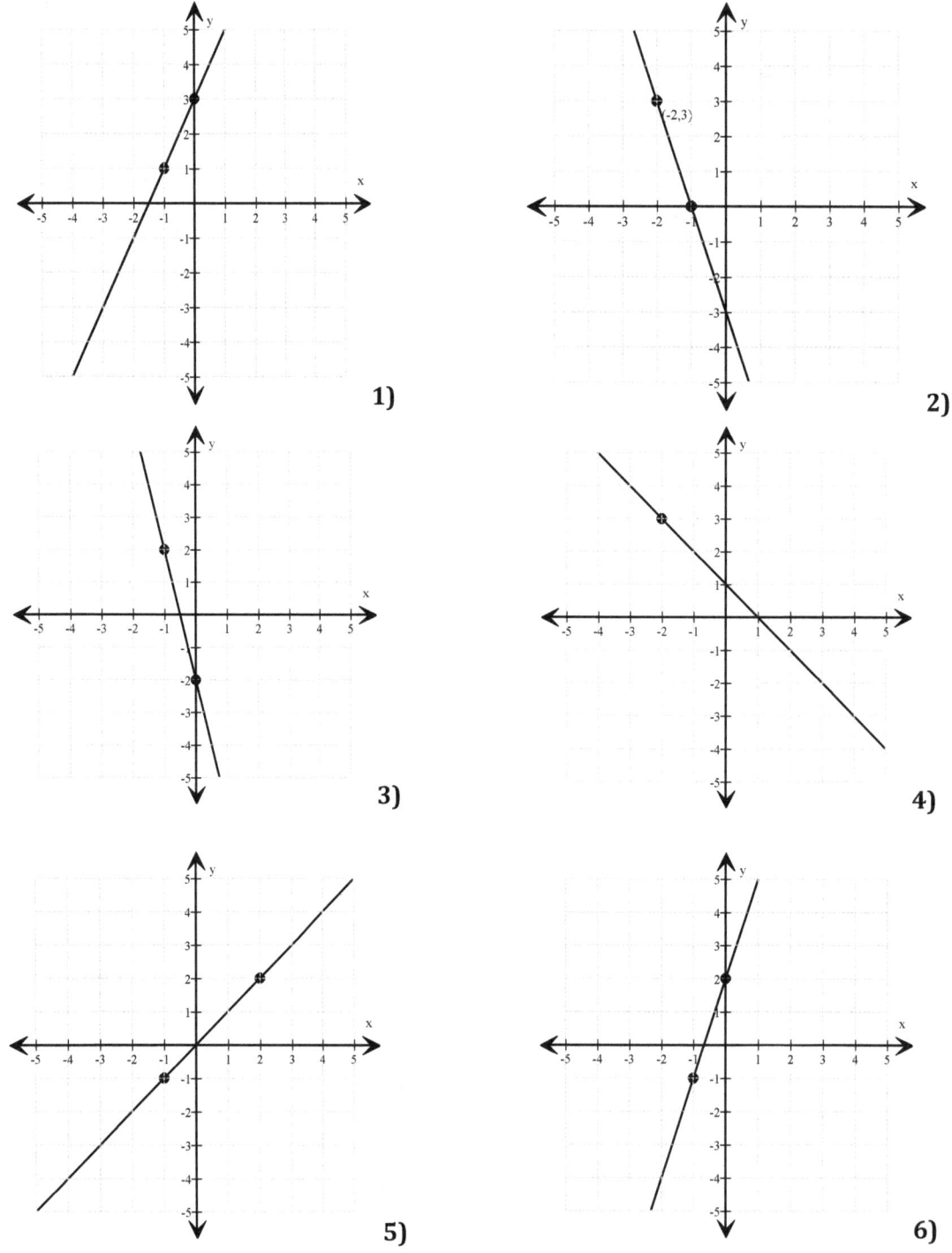

1)

2)

3)

4)

5)

6)

99) Different Types of Charts & Graphs

Graphs and charts make information clearer and easily understandable. Common graphs including bar charts, circle graphs, line graphs and picture graphs called also pictographs. Compare the four graphs. Which graph would you prefer?

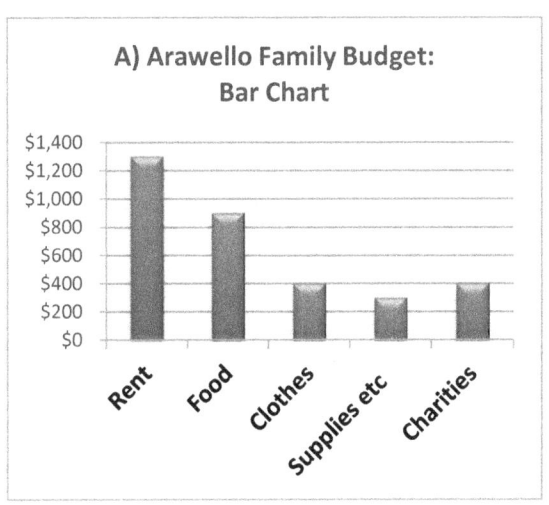

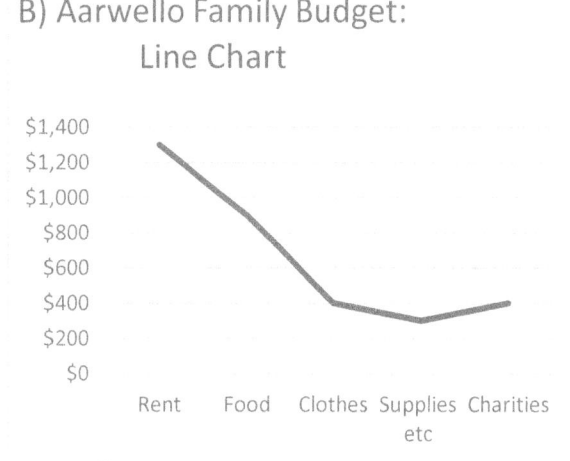

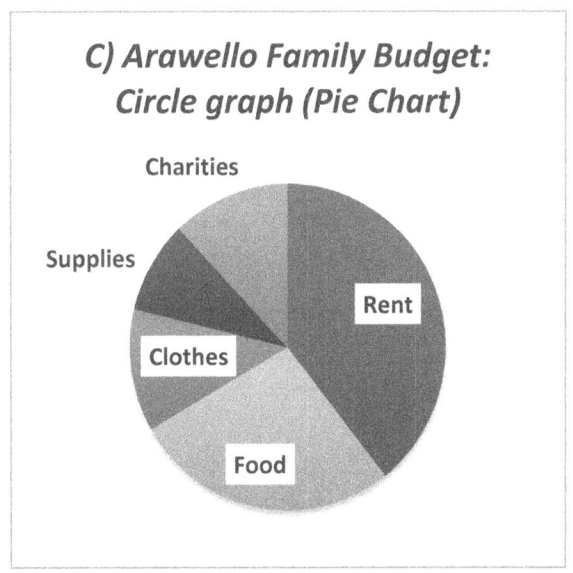

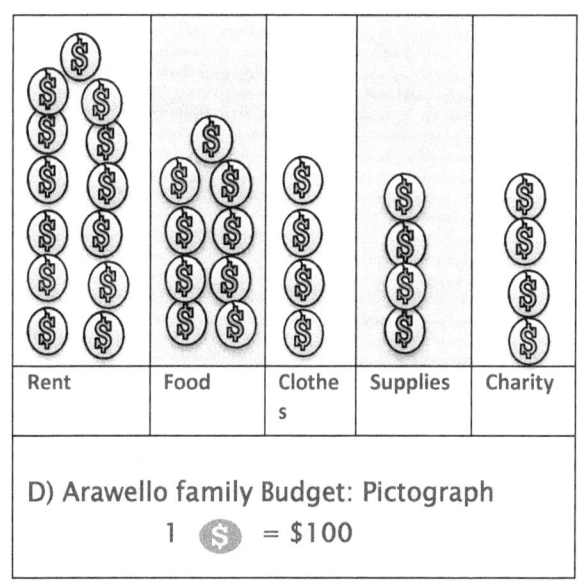

- Look at the title of the graph
- Understand the labels (food, weight, dollars etc.)
- Estimate the numbers if needed

E) Average smiles per day

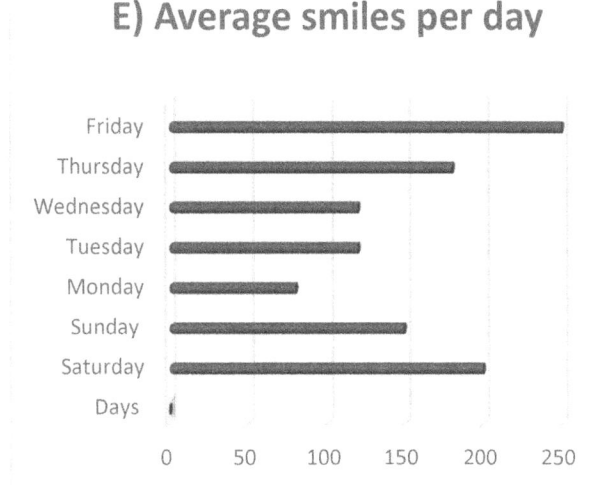

Use the information on graph (E) to answer questions 1 to 5.

1. Which day people smile most?
   _____ _____

2. Which day people smile the least?
   __ _____

3. Which two days have the same number of smiles?
   _____ and _____

4. How many more smiles do people have on Friday than Saturday?
   _____

5) How many times people smile on Friday, Saturday, and Sunday combined?

   __ __ _____

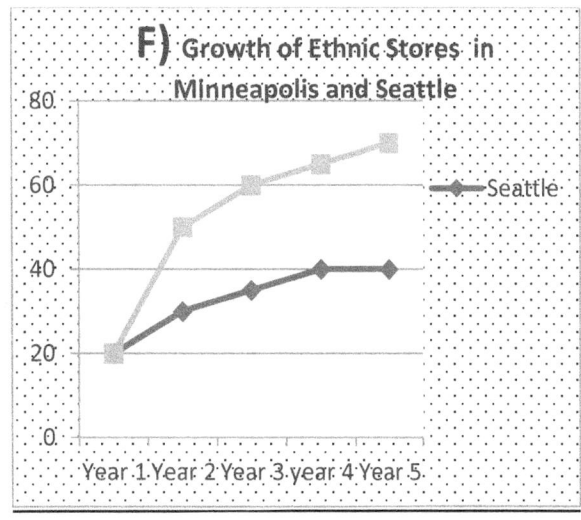

Use graph "F" on the left

6. How many more ethnic Stores were in Minneapolis in year 2 than in year 3?

   _____

7. Which year the two cities had same number of ethnic stores?

   _____

8) Total ethnic stores in year 4 in both cities: _____

9) How many ethnic stores were in Seattle in year 2 and year3?
   Year 2_____ Year 3_____

101) Geometry: Some Basic Definitions

A) What is Geometry?

Geometry is the study of the size, shape and position of figures. The word geometry itself is a Greek word and it means "earth measure"

Plane Geometry includes lines, triangles, circles and polygons drawn on a flat surface called plane.

Solid geometry includes three dimensional figures (3-D) such as cubes, cylinders cones, spheres, prisms, and pyramids.

Other types of geometry include analytic geometry, spherical geometry etc.

B) Basic Definitions

A Point has no size and no areas

A Line: Many points form a line that extends to both directions indefinitely.

A Line Segment has two endpoints.

A Ray goes only one direction and has one end point.

Parallel lines: Two lines that never cross each other.

Perpendicular lines: Two lines that cross each other at 90∘

Angle: Two line segments or rays that share a common endpoint form an angle.

Vertex: Where the line segments meet is called Vertex. "B" is the vertex.

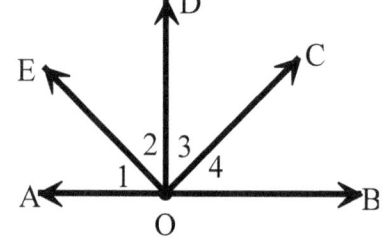

Angles are named by reading their labels: we can name the angles in three different ways:

∠ABC (read as Angle ABC) or ∠CBA or simply ∠B. You see the "B" is always in the middle in all cases.

Prcatice: Use the figure on the right: Name each angle in two ways

1) <1 _____

2) <2 _____

3) <3 _____

4) <4 _____

Name each angle indicated in three ways.

5)
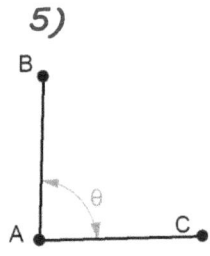

6)
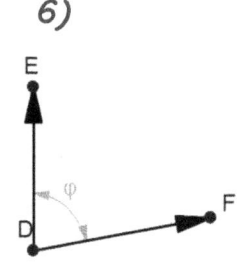

7)
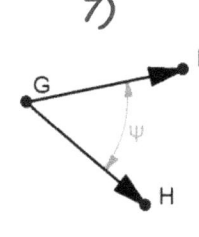

103) How to Measure Angles:

- ➤ Angles are measured usually through degrees (°).
- ➤ A full circle is 360 degrees (360°), A half circle is 180° and a quarter of a circle is 90°.

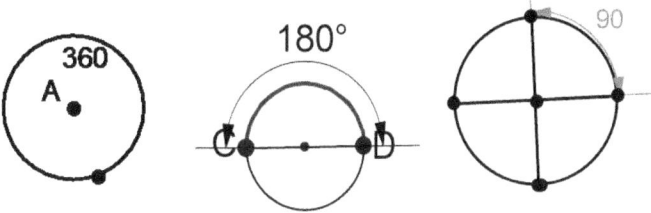

A **Protractor** is used to read the angle measurements.

- ➤ The vertex of the angle is lined up with the origin of the Protractor which is zero degree.
- ➤ But, the protractor has two scales: the inside scale and the outside scale.
- ➤ You have to know which one to read.

- o If the angle is more than 90° use the inner measure. It is 120° in the top protractor.
- o If the ,measure is less than 90°, use the inner angle. It is 60
- o To tell the measure of an angle, use $m \angle ABC$ which means "the measure of angle ABC?

Find the measures of these angles:

1) <PON =_____

2) <POM=_____

3) <POL=_____

4) <POK =_____

5) <KON=_____

6) <LOM=_____

7) <KOL= _____

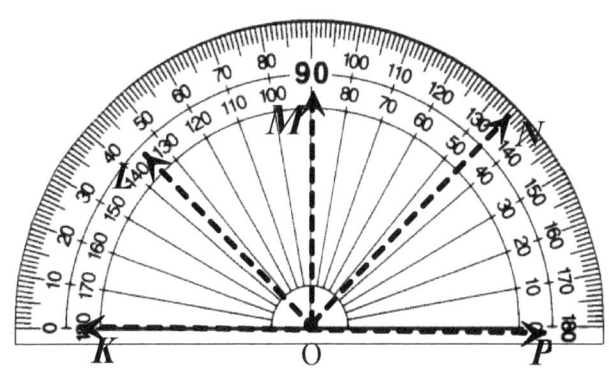

104) Classification of Angles

Angles are usually classified based on their measurements.

1) **Right Angle**: Two perpendicular lines make a right angle which is ninety degree **(90°).**

2) **Acute Angle**: is less than 90°.

3) **Obtuse Angle**: is greater than 90°

4) **A Straight Angle** is 180°

5) **Reflex Angle**: is an angle greater than 180 but less than 360

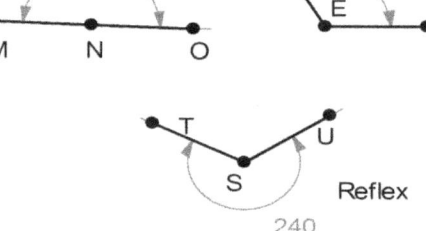

Classify the angles 1,2,3 as acute, obtuse or right. For problems 4,5 and 6, find the missing angles (x,y and z).

.1)

2)

3)

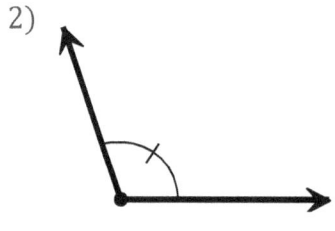

4)

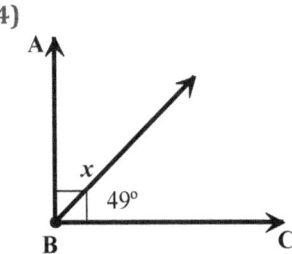

5)

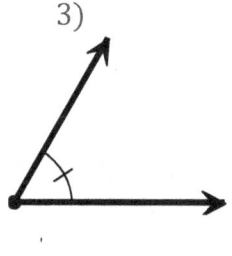

6)

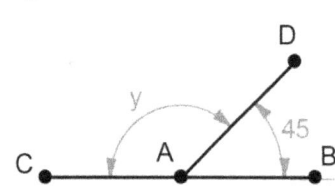

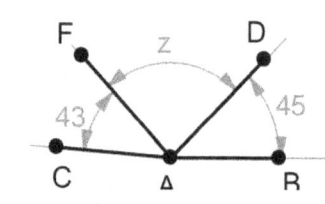

Complementary Angles: Two angles that add up to 90°.

Supplementary Angles: Two angles are complementary if their sum is 180

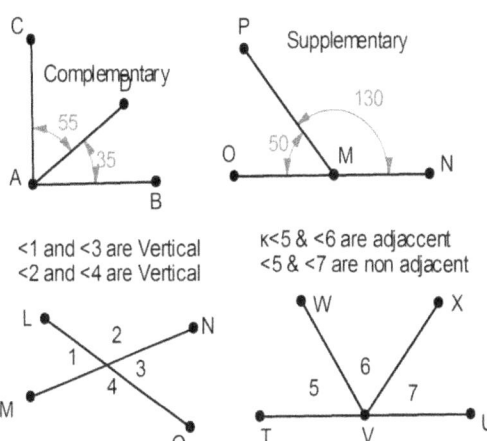

Vertical Angles: the opposite angles 1 and 3 as well as angles 2 and 4 are vertical angles Vertical anglesa are **congruent (equal)**

AdjacentAngles: <5 and <6 are next to each other and are called adjacent angles. Could you name another pair?

Alternate Exterior Angles: <1 and <8 are alternate exterior angles and are congruent Could you name another pair?

Alternate Interior Angles: <3 and <6 are alternate interior angles and are congruent Could you name another pair?

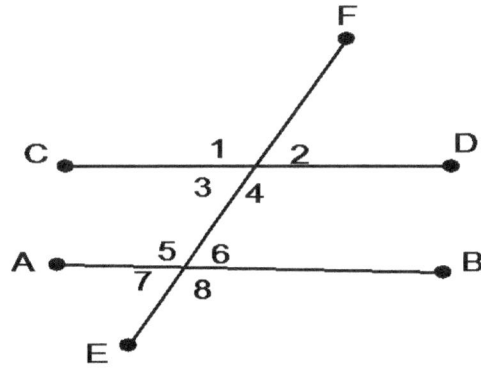

Corresponding Angles: <2 and <6 are corrresponding angles and are congruent. Could you name another pair?

106) Exercise in Angles

In the figure below, assume line l is a transversal. Identify three pairs of :

1) Vertical Angles

2) Adjacent Angles

3) Supplementary Angles

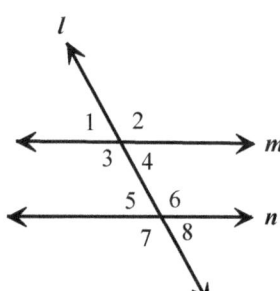

Use the figure below for problems 4 to 8

4) Name two interior alternate angles:

5) Name two exterior alternate angles

6) Name two vertical angles

7) Two corresponding angles

8) Name two supplementary angles

9) Find the measure of angle x.

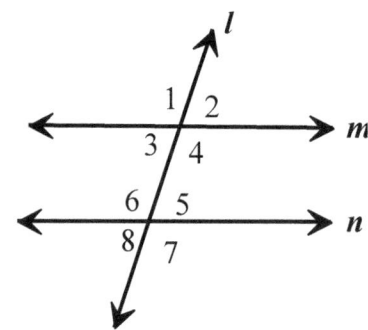

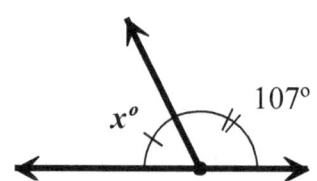

107) Triangles

Triangles: are figures that have three sides and three angles.

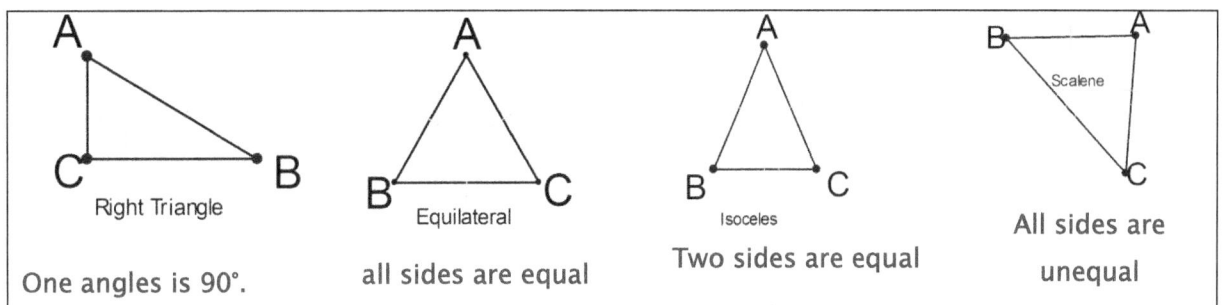

| Right Triangle | Equilateral | Isoceles | Scalene |
| One angles is 90°. | all sides are equal | Two sides are equal | All sides are unequal |

***Important: The sum of angles of any trianle is 180°.*

A) **Classify the following triangles** as equilateral, Isosceles or scalene: remember the marks show sides that are different or similar.

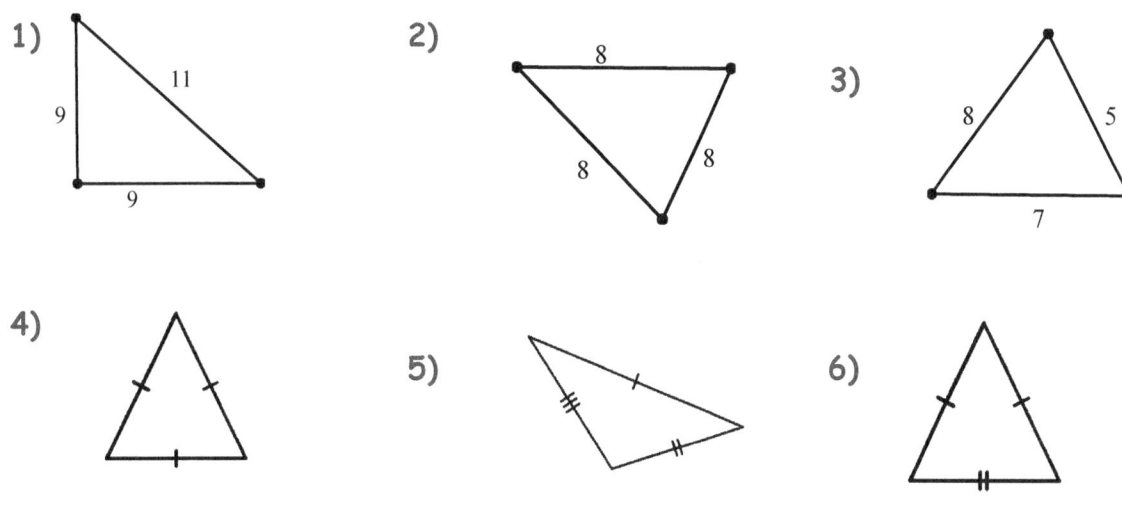

1) 9, 11, 9

2) 8, 8, 8

3) 8, 5, 7

4)

5)

6)

B) **Find the missing angle:**

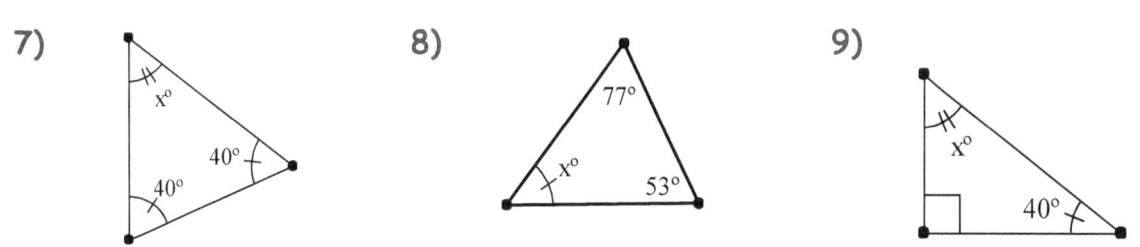

7) x°, 40°, 40°

8) 77°, x°, 53°

9) x°, 40°

108) The Pythagorean Theorem

The longest side of a triangle is called the **hypotenuse**. The two other sides are called

legs. In each right angle triangle, the squares of the legs is equal to the square of

hypotenuse.

$$hypotenuse\,^2 = leg1^2 + leg2^2$$

Mostly hypotenuse is shortened for c and
the legs as a, and b:

$$c^2 = a^2 + b^2$$

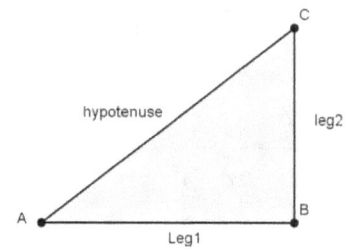

Example 1: Given a= 3, b=4 find the hypotenuse:	Example 2: Given c= 5, b=4 find the other leg of the triangle:
$$c^2 = a^2 + b^2$$ $$c^2 = 3^2 + 4^2 = 9 + 16 = 25$$ $$c = \sqrt{25} = 5$$	$$5^2 = a^2 + 4^2$$ $$5^2 - 4^2 = a^2$$ $$a^2 = 25 - 16 = 9$$ $$a = \sqrt{9} = 3$$

If c is the hypotenuse, find the missing sides. Use two decimal places as needed.

1) $a = 8$ $b = 6$ $c = ?$

2) $a = ?$ $b = 30$ $c = 40$

3) $a = ?$ $b = 6$ $c = 9$

4) $a = ?$ $b = 5$ $c = 8$

5) $a = 15$ $b?$ $c = 20$

6) $a = 36$ $b = ?$ $c = 49$

7) $a = 7$ $b = 4$ $c = ?$

8) $a = 8$ $b = 6,$ $c = ?$

109) The Distance Formula

Example: What is the length of segment (a) or the distance between points A and B?

The distance between any two points is calculated by using the distance formula

$$distance = \sqrt{(x_2 - x_1)^2 + (y_2 - y_1)^2}$$

$$d = \sqrt{(5-1)^2 + (4-1)^2}$$

$$d = \sqrt{16 + 9}$$

$$d = 5$$

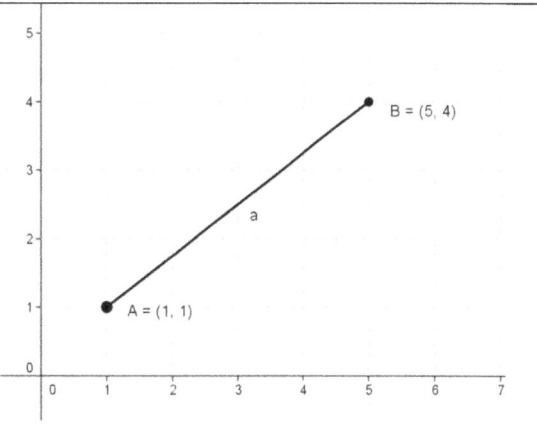

Calculate the distance of the following pairs of points. Use a calculator if you need to!

1. A (1,2), B(3,4)

2. C (4,5), D (6,4)

3. E (7,2), F(4,6)

4) G (-2,-3), H(7,-6)

5) I (-4,5), J (-8,8)

6) K(-3,-3), L(6,6)

7) M (-5, -6) N (2,4)

8) P (-6,4), Q (4,8)

9) R (5,10) S (-5,-6)

110) Special Triangles

Some triangles are so well known that they are called **Special Triangles.** It is easier to figure out the length of their sides and the measure of their angles.

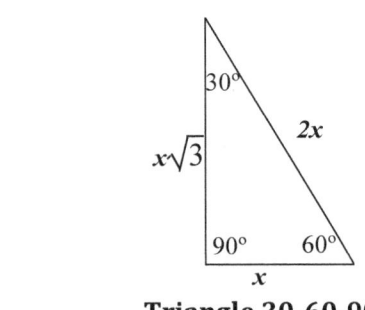 **Triangle 30-60-90**	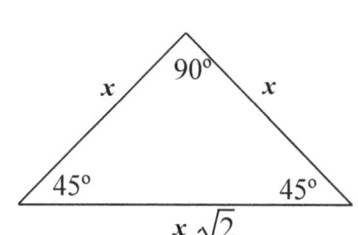 **Triangle 45-45- 90**
If the length of the side facing the 30 degrees angle is x, then the side facing the 90 degrees is 2x. The third side is x times radical three (x √3)	The length of the two sides facing the two 45 degrees are equal. To get the longest side multiply that number (x) by radical 2

Exercise: **find the missing measures: (x,y, p, q, r,s)**

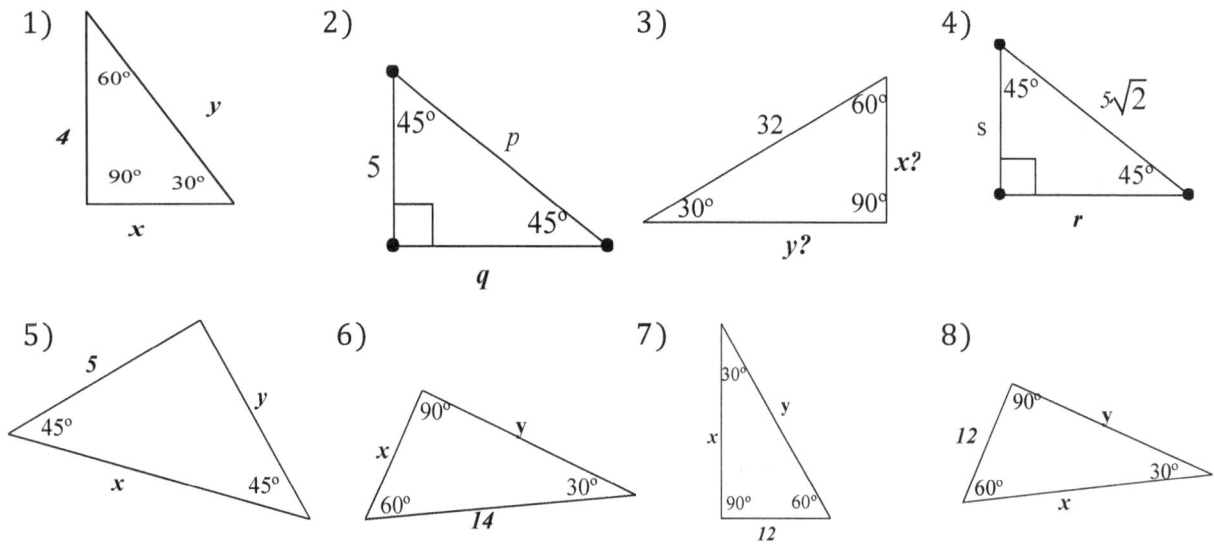

111) Quadrilaterals

Quadrilaterals: are figures that have four sides: quad means four and lateral means side.

1) Squares: have four equal sides and all of its angles are also right angles.

2.) Rectangles: have each two pairs of parallel lines equal. All angles are also right angles.

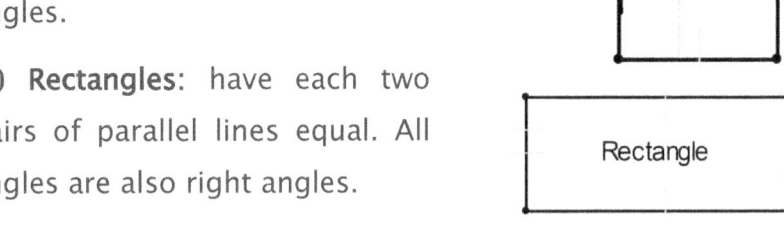

3) Trapezoid: has only one pair of parallel sides

4) Parallelogram: Each two opposite sides are equal. Opposite angles are also equal

5) Rhombus: The two pairs of opposite sides are parallel and all sides are equal.

6) Kite: the two pairs of adjacent sides are equal.

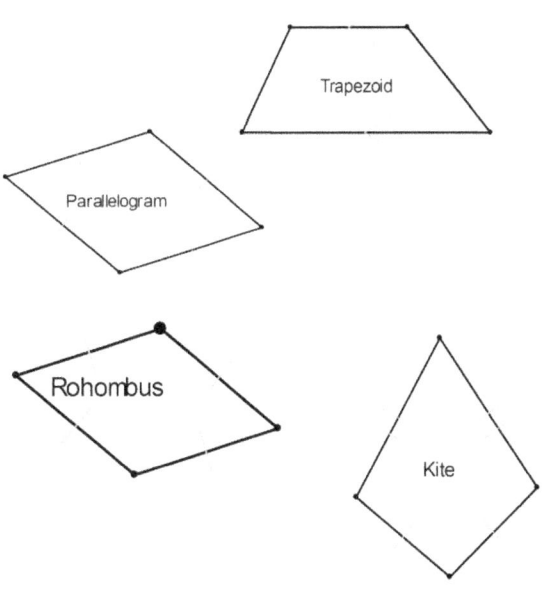

A **perimeter (P)** is the distance around the polygon. To find it, add the lengths of all sides.

The Area (A) of Polygons can be found in different ways (See Examples below)

Examples: Find the area and the perimeter for each of the following polygon (Assume all units are in feet).

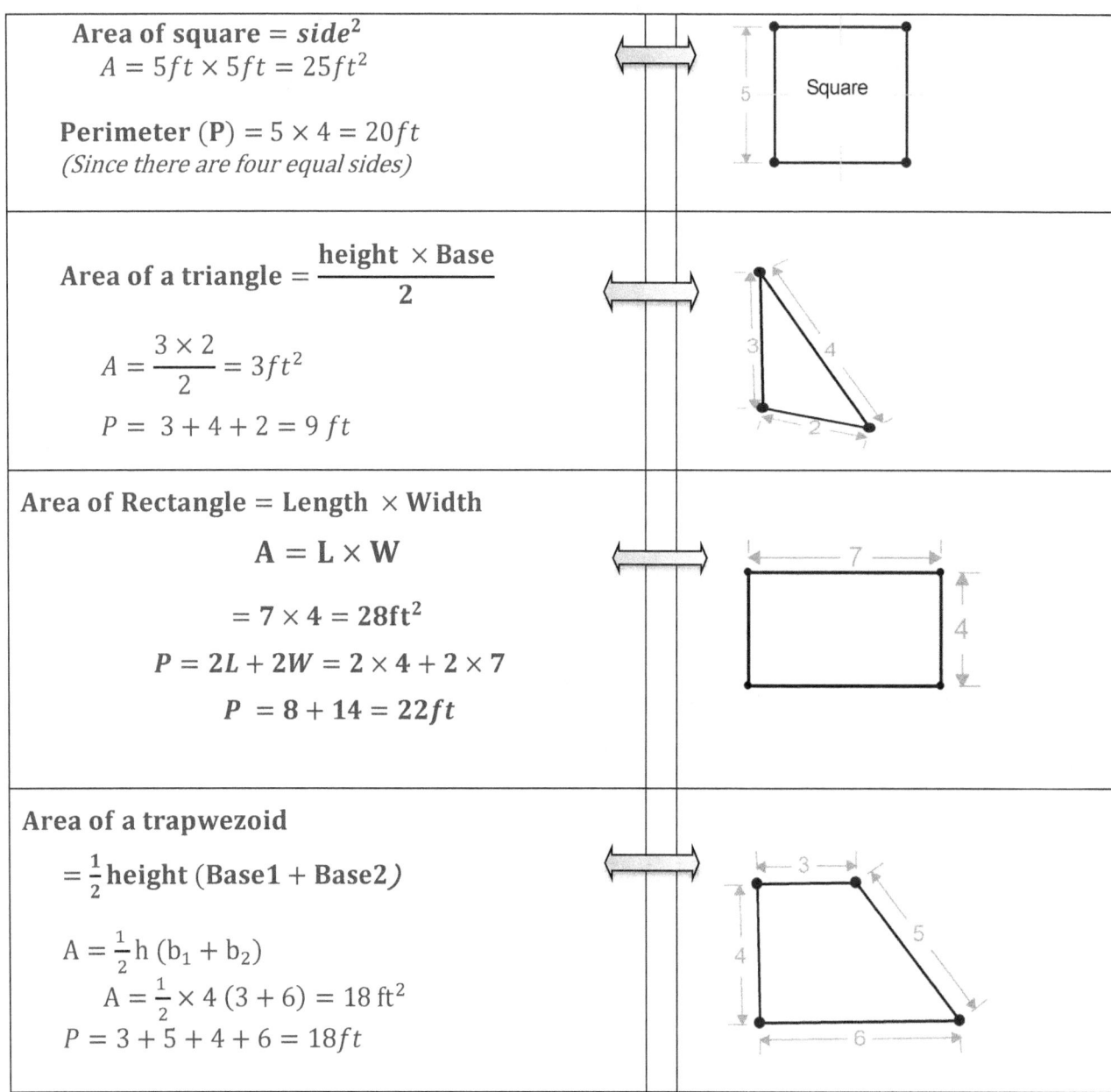

Area of square $= side^2$ $A = 5ft \times 5ft = 25ft^2$ **Perimeter (P)** $= 5 \times 4 = 20ft$ *(Since there are four equal sides)*	
Area of a triangle $= \dfrac{\text{height} \times \text{Base}}{2}$ $A = \dfrac{3 \times 2}{2} = 3ft^2$ $P = 3 + 4 + 2 = 9\ ft$	
Area of Rectangle = Length $\times$ Width $A = L \times W$ $= 7 \times 4 = 28ft^2$ $P = 2L + 2W = 2 \times 4 + 2 \times 7$ $P = 8 + 14 = 22ft$	
Area of a trapwezoid $= \frac{1}{2}\text{height (Base1 + Base2)}$ $A = \frac{1}{2}h\ (b_1 + b_2)$ $A = \frac{1}{2} \times 4\ (3 + 6) = 18\ ft^2$ $P = 3 + 5 + 4 + 6 = 18ft$	

Problems 1 to 9: Find the areas and the perimeters. Unless shown, assume all units are inches:

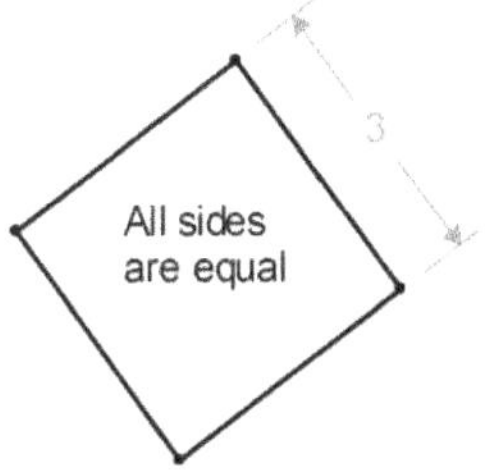

1)

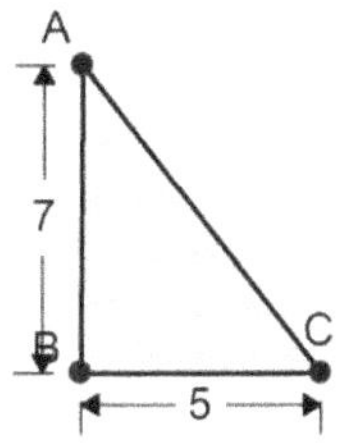

All sides are equal

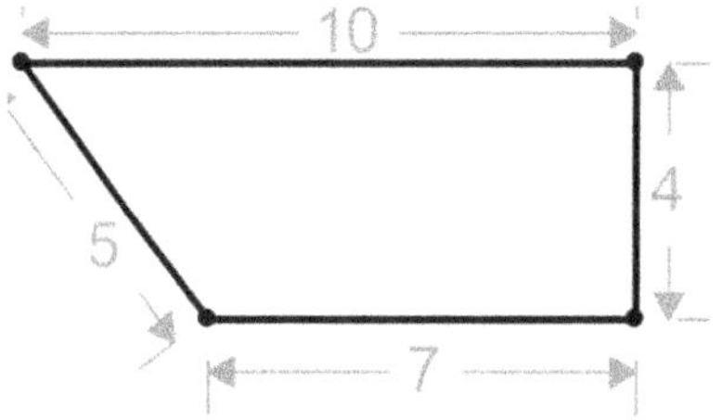

3

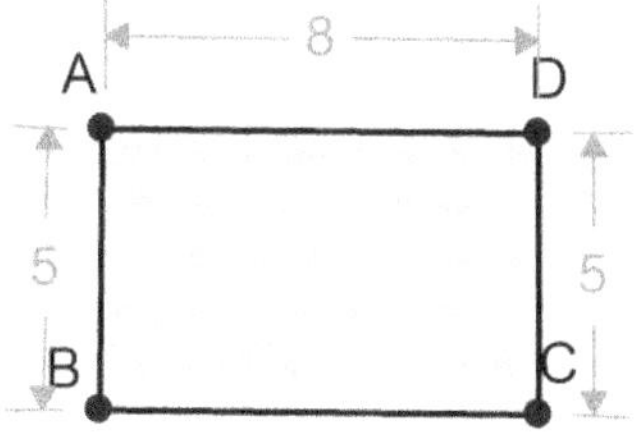

2)

A
7 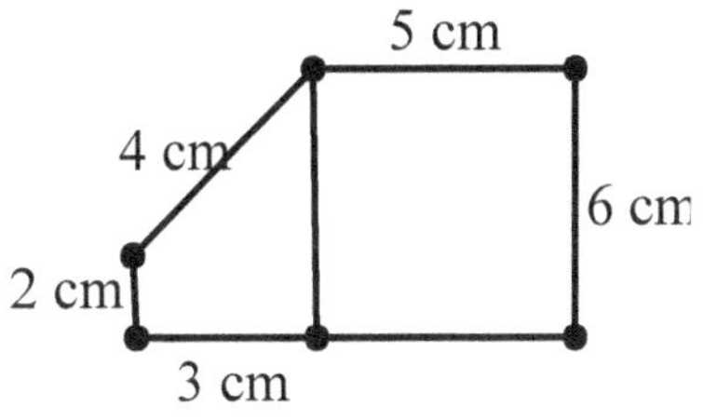
B
5
C

3)

10
5
4
7

4)

A
8
D
5
5
B
C

5)

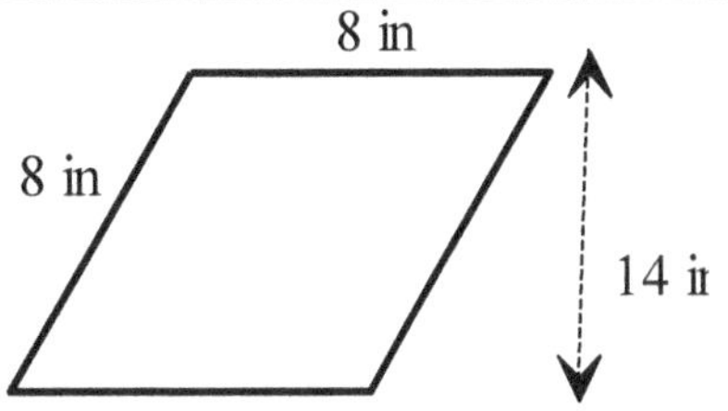

8 in
8 in
14 in

6)

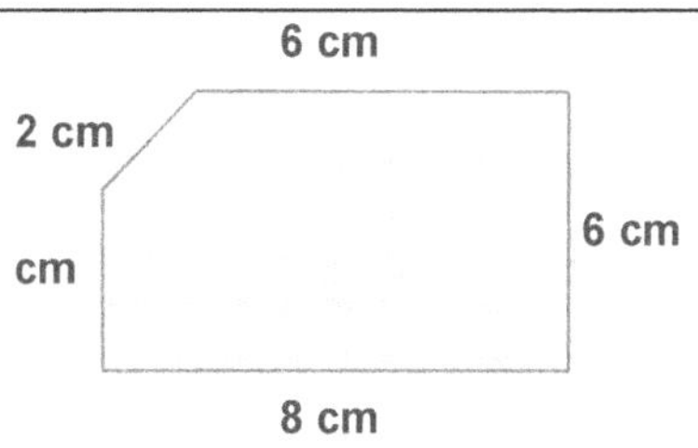

6 cm
2 cm
3 cm
6 cm
8 cm

7)

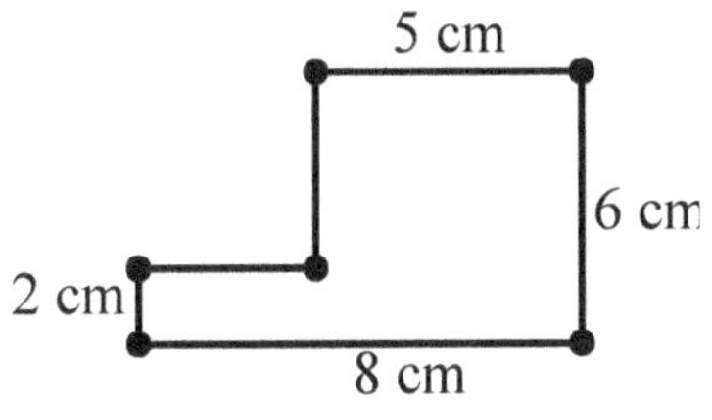

5 cm
6 cm
2 cm
8 cm

8)

5 cm
4 cm
6 cm
2 cm
3 cm

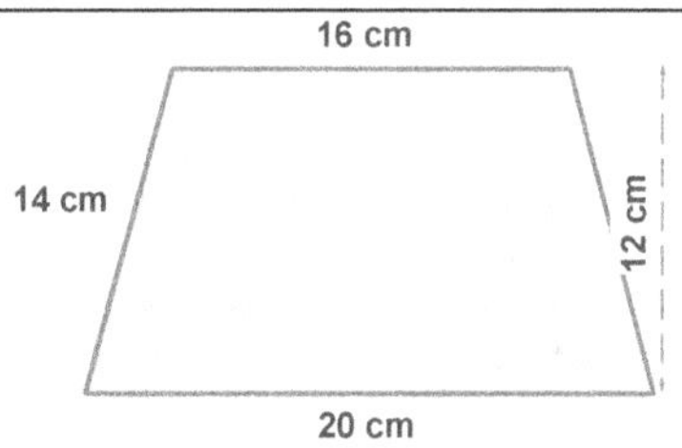

9)

16 cm
14 cm
12 cm
20 cm

The premeter is 18, find the length of the missing side HK?

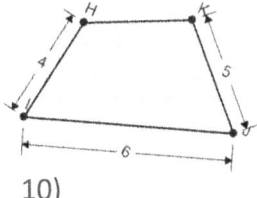

10)

(a) Find the area of triangle JHG
(b) Find the area of the trapezoid.

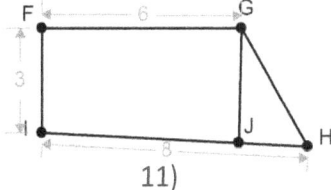

11)

123

114) The Circle

A) What is A Circle?

1. A circle is a shape that has all of its points the same distance from the center.

2. Circles are named after their centers. So, for example we have here Circle A and Circle C. (Below)

B) The Diameter and The Radius

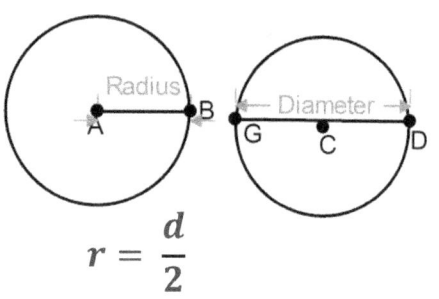

3. The Diameter (d) connects side to side of the circle passing through the center.

4. The Radius (r) is half of the diameter or the distance from the center to any points in the circle.

$$r = \frac{d}{2}$$

C) The Circumference and the Area:

5. The circumference (C) is like the perimeter: It is the distance around the circle.

$$C = 2\pi r$$

"r" is the radius and

$$\pi = \frac{22}{7} \text{ or } 3.14$$

6. The Area (A) of the circle is the area enclosed by the circle.

$$A = \pi r^2$$

D) Calculations: Example:

Calculate the diameter, the circumference and the area of a circle with a radius of 6 cm.

$$d = 2r = 2 \times 6 = 12 \, cm$$

$$C = \pi d = \frac{22}{7} \times 12cm = 37.7cm$$

$$A = \pi r^2 = \frac{22}{7} \times (6cm)^2 = 18.87cm^2$$

Find the circumference, and the area of the following circles:

1) $d = 4cm$ $C =?$ $A =?$

2) $r = 5\ cm$ $C = ?$ $A =?$

3) $d = 14\ cm$ $C =?$ $A =?$

4) $r = 7cm$ $C =?$ $A =?$

5) Name the Circles and then find the length of the diameter, the circumference and the area of each circle.

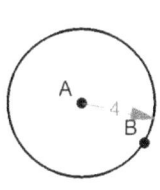 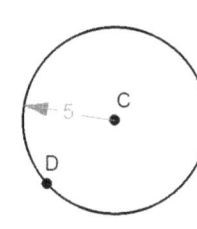 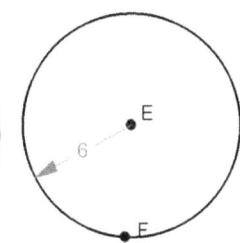

6) **True or False:** A circumference of a circle can be found by multiplying 3.14 with the diameter of the circle.

7) **True or False:** A circumference of a circle is π times its radius?

8) A pizza has a diameter of 8 inches. Find its (a) circumference and

   b) Its area.

True or False:

1. The sum of two supplementary angles is 90°.

2. Two adjacent sides are always 180°

3. All lines must be either perpendicular or parallel

4. All angles of a square are right angles.

5. A reflex angle is more than 180°

6) Opposite sides of a parallelogram are congruent.

8) Two intersecting lines can create vertical angles.

9) Corresponding angles can be different.

10) When both bases of a trapezoid are equal, the trapezoid becomes a rectangle.

11) Find the missing angles in each figure.

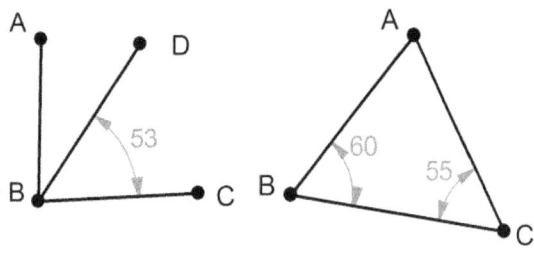

Problems 12-15:Find the perimeter and the area of the following figures:

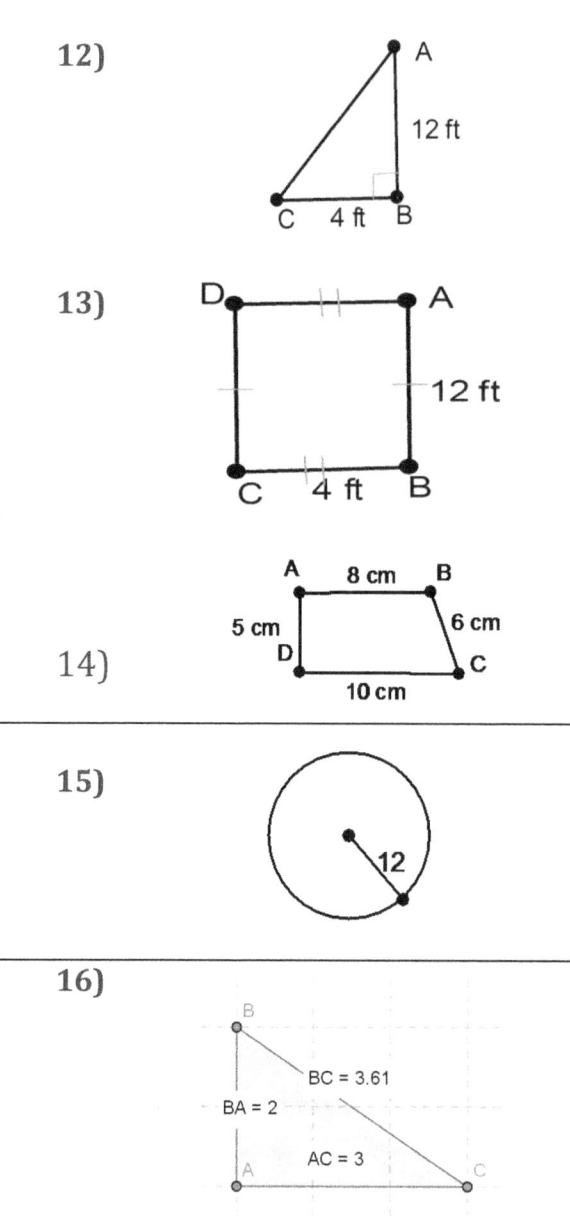

117) Finding Surface Area and Volume of a Cube

Example: Find (a) the surface area and (b) the volume of the cube that is 8 cm in each side

a) *Surface Area of a cube with side* $S = 6S^2$

   Surface area $= 6 \times 8^2 = 6 \times 64 = 384cm^3$

b) *Volume of a cube with side* $S = S^3$

   $Volume = 8^3 = 512cm^3$

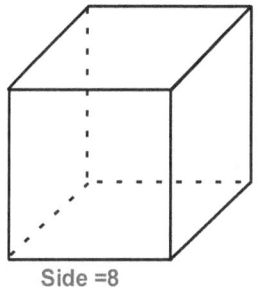

Side =8

Find the surface area of each of the following cubes with the given side. Show your calculations

1) S =5   2) S= 8   3) S= 10

4) S = 6   5) S= 12   6) S= 13

Find the surface area and the volume of each of the following cubes.

7) S = 11   8) S= 14   9) S= 9

118) Finding Surface Area and Volume of Rectangular Prism

Example : Find (a) the surface area and (b) the volume of a rectangular prism with the height of 3ft., length of 6 ft., and width of 4 ft.

a) $Surface\ Area = 2(h \times l + h \times w + l \times w)$
$surface\ Area = 2(3 \times 6 + 3 \times 4 + 6 \times 4)$
$$= 2(18 + 12 + 24)$$
$$= 2(54) = 108\ in^2$$

b) $Volume = l \times w \times h$
$$Volume = 6ft \times 4ft \times 3ft.$$
$$= 72ft^3$$

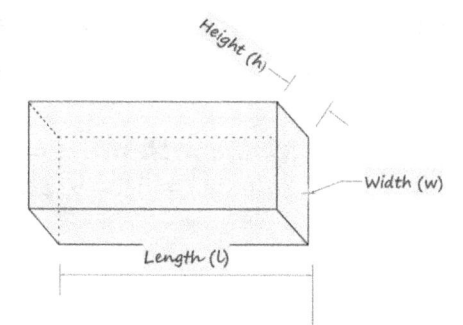

Find the surface area of a rectangular prism with the following dimensions. Make sure to round it to the nearest whole number.

1) Height = 6ft; length= 4ft; width = 5 ft.

2) Height = 4ft; length= 7ft; width = 3 ft.

3) Height =10ft; length =3ft; width = 2 ft.

4) Height = 12ft; length= 5ft; width = 5 ft.

Find the volume of a rectangular prism with the following dimensions:

5) Height = 6 cm; length= 4 cm; width = 5cm

6) Height = 12 cm; length= 10 cm; width = 8 cm

7) Height = 9 cm; length= 3 cm; width = 12cm

8) Height = 11 cm; length= 9cm; width = 10 cm

a) $\textbf{Surface Area} = 4\pi r^2$
$$= 4 \times 3.14 \times 5^2 = 314cm^2$$

b) $\textbf{Volume} = \dfrac{4}{3}\pi r^3 =$
$$= \dfrac{4}{3} \times 3.14 \times 5^3 = 523.3\ cm^3$$

radius (r)

Find the surface area of each of the following sphere with the given radius in cm. Round your answer to the nearest whole number.

1) r =3

2) r = 8

3) r = 10

4) r = 7

5) r = 11

6) r = 12

Find the volume of each of the following spheres with the given radius. Show your calculations

7) r = 14

8) r = 20

9) r = 15

120) Find Surface Area and Volume of a Cylinder

Example : Find (a) the surface area and (b) the volume of a **cylinder** with radius of 3 cm and height of 10 cm.

a) **Surface Area** $= 2\pi r^2 + 2\pi rh$

$$= 2 \times 3.14 \times 3^2 + 2 \times 3.14 \times 3 \times 10$$

$$= 56.52 + 188.4 = 244.92 cm^2$$

b) **Volume** $= \pi r^2 h = 3.14 \times 3^2 \times 10 = 282.6\ cm^3$

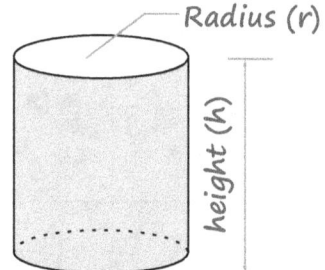

Find the surface area of each of the following cylinder with the given radius and height. The dimensions are all in meters (m). Round your answer to the nearest whole number.

1) r =3; h= 8

2) r = 4; h= 10

3) r = 10; h = 20

4) r = 7; h= 12

5) r = 11; h = 21

6) r = 12; h= 16

Find the volume of each of the following cylinder with the given radius and height. The dimensions are all in meters (m). Round your answer to the nearest whole number.

7) r = 2 ; h= 8

8) r = 6; h= 30

9) r = 8; h= 12

121) Find Surface Area and Volume of a Cone

Example: Find (a) the surface area and (b) the volume of a cone with slant length of 10 ft. , height of 6 ft. and radius of 5 ft.

a) Surface Area $= \pi rs + \pi r^2$

$\qquad = 3.14 \times 5 \times 10 + 3.14 \times 5^2$

$\qquad = 157 + 78.5$

$\qquad = 235.5 \; cm^2$

b) Volume $= \dfrac{1}{3} \pi r^2 h$

$\qquad = \dfrac{1}{3} \times 3.14 \times 5^2 \times 6$

$\qquad = 157$

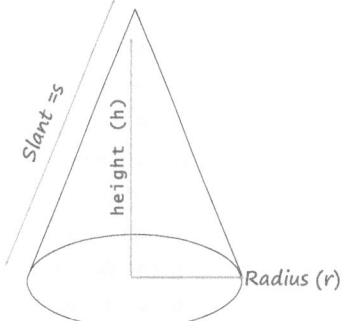

Find the surface area of a cone with the given dimensions. All dimensions are in ft.

1) r =3; s = 8

2) r = 4; s = 10

3) r = 5; s = 20

4) r = 7; s = 12

5) r = 6; s = 21

6) r = 8 ; s = 16

Find (a) the surface area and (b) the volume of a cone with slant length of 10 ft. Height of 6 ft. and radius of 5 ft.

7) r = 2 ; h= 8;
   s = 14

8) r = 6; h= 30
   s = 16

9) r = 8; h= 12
   s = 20

122) Finding Surface Area and Volume of Pyramid

Find the surface area and volume of the pyramid shown:

Assume base =5 cm ;   Slant =10 cm;   Height= 8 cm

Surface Area =

$$\textbf{Base area} + \frac{\textbf{perimeter of base} \times \textbf{slant height}}{2}$$

$base\ area = 5 \times 5 = 25$

$Perimeter\ = 4 \times 5 = 20$ (four sided)

$$Surface\ Area = 25 + \frac{20 \times 10}{2} = 125\ cm^2$$

b) $Volume = \dfrac{1}{3} \times base\ area \times height$

$$= \tfrac{1}{3} \times 25 \times 8 = 66.7 cm^3$$

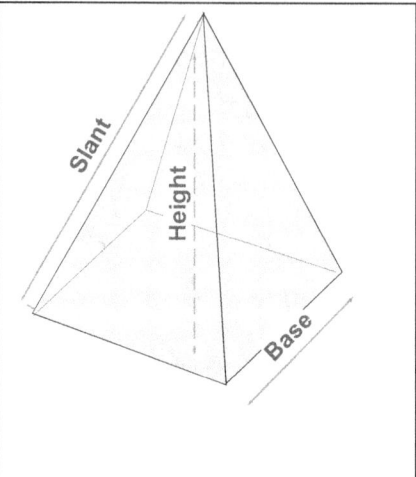

Find the surface area of the square pyramid with the given slant and base. Use 3.14 for π.

1) base = 4 cm;   slant = 10 cm

2) base = 3 cm;   slant = 9 cm

3) base = 4 cm;   slant = 9 cm

Find the Volume of the square pyramid with the given slant, base and height. Use 3.14 for π.

4) base = 5 cm;   slant = 10 cm;   height =8 cm.

5) base = 3 cm;   slant = 9 cm;   height =8 cm.

6) base = 4 cm;   slant = 9 cm;   height =8 cm.

123) Review of three Dimensional Area and Volumes

Find the surface area of the following figures with the given dimensions:

1) **Cube**: Side length = 15 cm.

2) **Rectangular prism:** height = 12ft; length = 6ft; width = 4ft

3) **Sphere**: radius of 11 cm.

4) **Cylinder**: radius of 10 m and a height of 15 m.

5) **Cone** with a radius of 10 cm and slant of 14 cm.

6) **Square pyramid** with a slant of 12 m and a base of 5m.

Find the Volume of the following figures with the given dimensions:

7) **Cube**: Side length = 15 cm.

8) **Rectangular prism:** height = 12ft; length = 6ft; width = 4ft

9) **Sphere**: radius of 10 cm.

10) **Cylinder**: radius of 12 m and a height of 15 m.

11) **Cone** with a radius of 9 cm, slant of 14 cm and a height of 11 cm.

12) **Square pyramid** with a slant of 11m and a base of 5m, and a height of 10 cm.

124. Statistical measures: Introduction

Teachers usually **average** your test scores to post for you one final score. They talk about mean score. **The mean** doesn't mean a mean person etc. but just an average of, for example, your three test scores. Other common statistical measures are the **medium, the mode and the range.** Remember in general, Statistics has three main goals: collecting, summarizing, and analyzing data

Example:

Omar took seven tests and scored these grades: **calculate the mean, the medium and the mode.**

Test 1	Test2	Test3	Test 4	Test5	Test6	Test7
85	83	79	90	90	80	88

1. **The mean:** $(\bar{x})$ is the average: adding all scores and dividing by the number of tests.

$$\bar{x} = \frac{85 + 83 + 79 + 90 + 90 + 80 + 88}{7} = \frac{595}{7} = 85$$

2. **The medium:** is the middle number in the data. Write data in increasing or decreasing order.

$$79 \quad 80 \quad 83 \quad \mathbf{85} \quad 88 \quad 90 \quad 90$$

The middle score become 85 again but it is not always equal to the mean!
Note: If the data is even you have two middle scores and you have to average the two.

3) **The mode:** is the number that shows up most. In our scores 90 shows up twice. Therefore, the mode is **90**. If two different numbers show up at the same frequency, we say the data is bimodal (means has two modes).

4) **The Range:** Is the difference between the highest score and the lowest score.
$$95 - 65 = 30$$

125) Statistics Problems 1:

1. Aisha took 5 math tests and scored: 90, 95, 80, 85 and 95.
 Calculate her mean, medium, mode and range of her scores.

2. Abigail scored 90, 95, 80, 85 95, and 75 in her math 6 scores. Calculate her mean, medium, mode and range of here scores.

3) To receive an "A" in math, Mustafa has to score a mean grade of 90% in his 5 tests. He scored 85, 90, 87, 88 in his first four scores. How much should he score in the fifth score to get an "A"?

4) In five games in 2014 against Milwaukee, Indiana, Washington, Orlando, and Toronto, Le Bron James scored 26, 19, 29,29, 15 He played 41, 32, 36, 31, and 37 minutes respectively.

     a) What is his mean, medium, mode and range scores?

     b) What is his mean and medium minutes of play in each game?

126) The Five Number Summary

Another way to summarize data is to calculate the **Five Number Summary** which are: **Minimum** (lowest score), **Maximum** (the highest score), **Medium** (middle score), **Quartile one (Q1)** and **Quartile three (Q3)**. Let's use Omar's test scores again:

Omar took seven tests and scored these grades: **calculate the five number summary.**

Test 1	Test2	Test3	Test 4	Test5	Test6	Test7
85	83	79	90	90	80	88

1. **The Minimum = 79**
2. **The medium or the mid number** after ranking the data **is 85:**

    79, 80, 83, **85,** 88, 90, 90

3. **Quartile 1 (Q1)** is the midpoint between the minimum and the medium. Since there are even numbers (80, and 83) we need the average of the two: **(80+83)/2= 81**
4. **Quartile 3 (Q3)** is the midpoint between the medium and the maximum. Since there are even numbers (88, and 90) we need the average of the two: **(88+90)/2= 89**
5. **The Maximum = 90**

The box and whisker plot is simple a visual or graphical way to show the five number summary

---

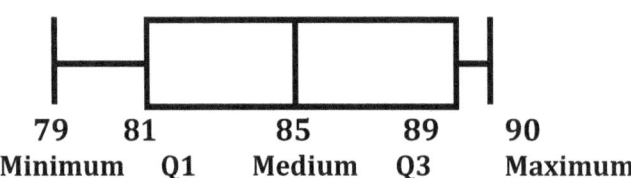

79	81	85	89	90
Minimum	**Q1**	**Medium**	**Q3**	**Maximum**

1) Use the following weights in pounds of a typical student weights to answer the questions below:

160, 145, 205, 150, 157, 138, 145, 210, 135, 190

a) Calculate the five number summaries of the weights.

b) Show the whisker and plot box of the data

2) Gulled visited the gym all seven days of the last week and exercised the following minutes:

60, 50, 45, 42, 55, 47, 35

a) Calculate the five number summaries.

b) Sketch the box and whisker plot.

127) Welcome to Probability

Probability is the likelihood for an event to happen. What is the chance or the likelihood that it will snow tomorrow in Minneapolis? What is the likelihood to get correct answer in my multiple choice questions if I just guess? What is the probability (the likelihood) that you will be the president of the United States?

Probability is used in biology, chemistry, physics, social sciences, insurance, and many other fields. For example it is used to predict the likely winners and losers in a presidential election. Let's follow the chart below to guide our discussion:

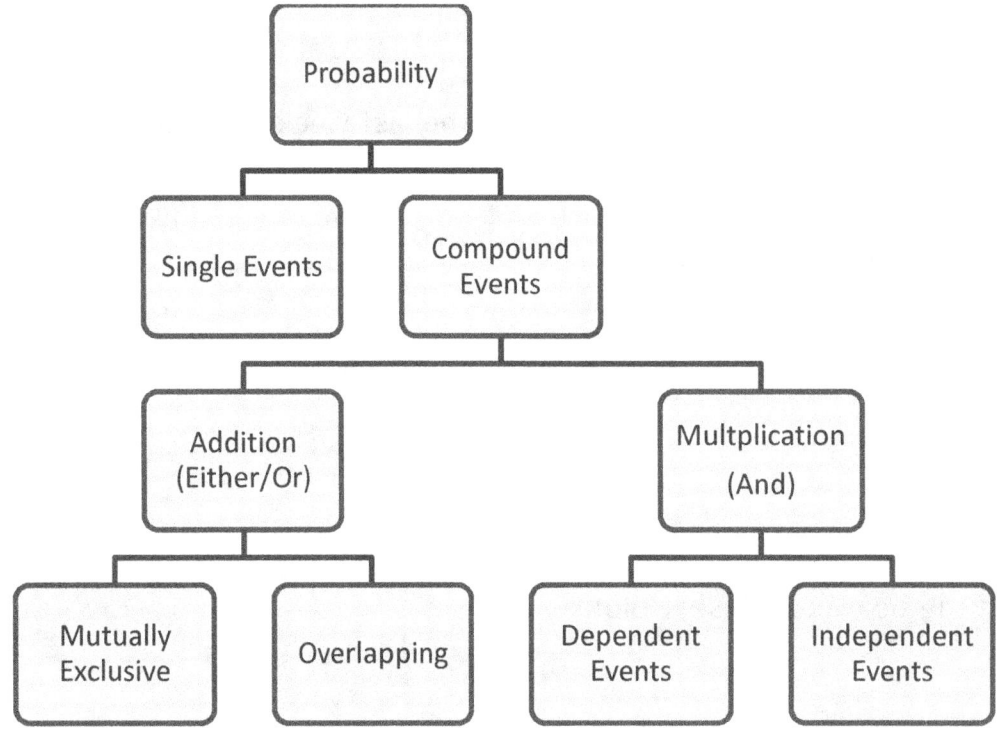

128) Probability of Single Events

An event is the outcome that we are predicating or interested in. <u>Single Event</u> is just one event like throwing a coin or die once

> **The probability of an event** is the ratio of the number of favorable outcomes (called also successes) of an event over the total number of possible outcomes. Favorable outcome means the number of times that the event you are testing will happen.

> $$P = \frac{\text{Number of favorable outcomes}}{\text{Total number of Possible Outcomes}}$$

>

> Also, the probability of any event is always a fraction or a decimal between 0 and 1.
> $$0 \leq p \leq 1$$

> **A Probability of 1 means we are 100% sure. A zero probability event is an impossible event.**

Example 1: A coin is flipped in the air. What is the likelihood that it will land on its head?

A coin has 2 possible outcomes: Head or Tail

Here there is only one head. The number of successes is 1

$$P = \frac{1}{2}$$

Example 2: A die is tossed once. What is the probability of getting a number greater than 2?

A die has six possible outcomes since it has 6 sides.

All four numbers (3, 4, 5, and 6 are more than 2).

⇨The number of successes= 4

$$p = \frac{4}{6} = \frac{2}{3}$$

Use the data for problems 1 to 10: a bag contains 6 different numbers: 2, 4, 6, 8, 10, and 20. If person picks just one number:

1. What is the probability that the number is ten?

2. What is the probability that the number is less than 6?

3. What is the probability that the number is less than 2?

4. What is the probability that it is more than 20?

5. What is the probability that it is a number more than 4?

6. What is the probability that the number is 4 or more?

7. What is the probability that it is greater than 10?

8. What is the probability that that number is divisible by 5?

9. What is the probability that the number is divisible by 10?

10.   What is the probability that the number is 25?

Use this data for problems 11–13: Find the probability that the spinner:

11) Will land on black

12) Will land on white?

13) Will land on yellow or black?

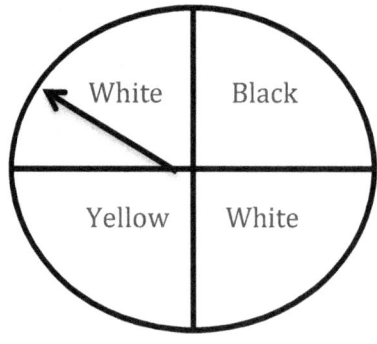

129) Probability of Compound Events

Compound Events: are two or more events happening at the same time, like tossing a coin more than once, or two coins (or more) at the same time.

Example1: What is the probability of getting three heads when you toss 3 coins once each?

Each coin has one out of two chances of landing on its head: Therefore, for all three coins:

$$P(3 \text{ heads}) = \frac{1}{2} \times \frac{1}{2} \times \frac{1}{2} = \frac{1}{8}$$

Example2: What is the probability of two dices both landing on 5 at the same time?

Since each dice has 6 sides, each dice has one out of 6 chances in showing the number 5: Therefore, for all three coins:

$$\mathbf{P(5)} = \frac{1}{6} \times \frac{1}{6} = \frac{1}{36}$$

Practice Problems:

Problem 1: What is the probability of getting four heads when you toss 4 coins once each?

Problem 2: What is the probability of getting of all tails when you toss a coin five times?

Problem 3: What is the probability of three dices all landing on 1 at the same time?

Problem 4: What is the probability of three dices all landing on 6 at the same time?

130) Either/OR (Addition): Mutually Exclusive Events

Sometimes we want to know **if one event or another** will happen. There are two kinds of such events: (a) Mutually exclusive and (b) Non-mutually exclusive

---

Mutually exclusive Events are two events where happening of either one excludes the other to happen. That means if one happens the other can't happen. A coin landing on a head and tail are two mutually exclusive events at any trial

$$P(A \text{ or } B) = P(A) + P(B)$$

Example: Your classmates have organized a fund raising event in which there will be one winner. The class sold 50 tickets. Saida bought 3 tickets and her friend Muna bought 2 tickets.

   a) Is this a mutually or non-mutually exclusive events?

   Yes. See that only one winner will be elected. If Saida wins, it is impossible for Muna to win

   *b) What is the probability that **either Saida or Muna** win?*

   $P \text{ (elect Saida)} = \frac{3}{50}$ (She has 3 tickets)

   $P \text{ (elect Muna)} = \frac{2}{50}$ (she has 2 tickets)

$$P(S \text{ or } M \text{ wins}) = P\ (S \text{ wins}) + P(M \text{ wins}) = \frac{3}{50} + \frac{2}{50} = \frac{5}{50} = \frac{1}{10}$$

Your Practice Problems:

There are 5 green marbles, 4 red marbles and 3 blue marbles. (a) What is the probability of picking (a) a green marble? (b) a green or blue marble? (c) a green, a blue or a red marble?

> **Non-Mutually Exclusive Events occurs** when both events can happen at the same time. Drawing from a deck of cards **a 6 or a red card** is dependent because you can pick up one card that is both red and is also 6. The formula for Dependent events is:
>
> $$P(A \text{ or } B) = P(A) + P(B) - P(A \text{ and } B)$$

Example: If one card is drawn from a deck of card, what is the probability that it is a king and black? (For those not familiar with cards, there are 52 cards in a deck: 26 black and 26 red).

Solution:

We can pick up one card that is both a **king** and a black card. So, the event is Non-exclusive. So, P (K or B) = P (K) + P (B) – P (K & B)

$P(K) = \dfrac{4}{52}$ (There are 4 kings in the 52 cards); $P(B) = \dfrac{26}{52}$ (There are 26 black cards)

$$P(K \text{ or } B) = \dfrac{4}{52} + \dfrac{26}{52} - \dfrac{2}{52} = \dfrac{28}{52} = \dfrac{7}{13}$$

<u>Useful Hint</u>: (It is good practice to reduce fractions only at the end)

Practice Problems: If you are picking just one card from a deck of cards, what is the probability of picking up?

   (a) a red queen; (Hint there are just 2 red queens in the cards)

   (b) a queen or red card? (hint: there are four queens and 26 red cards including the 2 red queens).

2) If you choosing just one card from a deck of cards, what is the probability of picking up: (a) a black jack; (b) a black jack or diamond. (There are 13 diamonds in the deck of cards.)

132) And Events (Multiplication): Independent Events

Sometimes, we want to know the likelihood that two events will both happen (or probability of A and B). There are two kinds of such events: Independent events and dependent events:

Independent Events: When the outcome of one event doesn't affect the outcome of the other event. The general formula for such events is:

$$P \text{ (A and B)} = P(A) \times P(B)$$

Example: In a recent Poll 1/4 of the students in the school said they like math and 2/3 of the students they watch movies on weekends. What is the probability that a randomly chosen student will love math and watch movie?

Solution: *These two are independent events.*

P(math and movie) = P(math) × P (movie).

$$P(\text{like Math}) = \frac{1}{4} \text{ (given)} \qquad\qquad P(\text{watch movie}) = \frac{2}{3} \text{ (given)}$$

$$P(math \ \& \ movie) = \frac{1}{4} \times \frac{2}{3} = \frac{2}{12} = \frac{1}{6}$$

You Practice Problem:

Three out of four PreU students love math. ¼ play basketball. What is the probability that a randomly chosen student will love math and play basketball?

B) **Dependent Events**: When the outcome of one event affects the outcome of the other event. The general formula for such event is:

$$\textbf{P (A and B)} = \textbf{P (A)} \times \textbf{P(B given A)}$$

The Probability of **B** given **A** means you should account for how the happening of event A has affected the happening of event B.

Example:

In a deck of 52 playing cards what is the probability of drawing a queen first and then a queen in the second draw **without replacement** (without putting back what you get first).

These two dependent events. P (A and B,) = P(A)×P(B given A)

$P(\text{Queen}1) = \frac{4}{52}$ (there are four queens in 52 cards)

$P(\text{Queen } 2) = \frac{3}{51} =$ (We have subtracted the queen drawn in event = The number of cards decreased also by one)

$P(Q1 \text{ and } Q2) = \frac{4}{52} \times \frac{3}{51} = \frac{1}{221}$

Comment: The chance of getting 2 queens is very small!

Your Practice Problem:

In a deck of 52 playing cards, what is the probability of drawing (a) a king first and a queen second? (a) a king first and a king second? (c) a red jack first and a red jack second? (d) a red jack first and a black queen second?

(Assume no replacements)

1. If you roll a die only one time, what is the chance that it will land on 3 or 4?

2. You just got a new credit card. What is the probability that the last number is (a) 9; (b) 5; or (c) odd number?

3. *If you toss a dice three times what is the probability of getting 3, 4, or 6?*

4. What is the probability of drawing a red ball from a basket containing four balls with different colors (green, blue, red, and yellow)?

5. Saynab has *five scarves with different colors in her purse. The colors are green, blue, purple, orange and red. What is the probability that (a) she picks yellow? (b) That she picks green*

6. *What is the probability that a spinner arrow will stop at an :*

   (a) 8 or 1?

7. *(b) odd number*

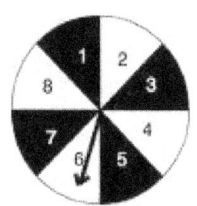

135) Measurements 1: The basic units

Who wears the largest shoes in the NBA players? Well, It is said that LeBron James wears size 14 (inches). But, Shaquille O'Neal', wears size 15 (inches) (Source: NBA.com)

So, if there were no units of measure for shoes, milk, clothes, buyers and sellers would be very confused. Sure, Shaq would have gone from store to store and try the right size. That would be a lot of work!

The U.S measures and the Metric System

There are two systems of measurements: The **Metric Measures** used in most of the world and **U.S Measures** used mainly in the U.S and Canada. Listed below are most common units of length, weight, and volume. **Also don't forget the abbreviations.** Math people use these abbreviations often.

The U.S measure		The Metric Measures
inches (in) feet(ft) yards (yd) miles (mi)	**Length**	millimeter (mm) centimeter (cm) meter (m) kilometer (km)
ounces (oz) pounds (lb), tons (t),	**Weight**	milligram(mg) centigram (cg) gram (g) kilogram (kg)
fluidounces (fl oz) cups (c) pints (pt) quarts (qt), gallons (gal)	**Volume**	milliliters (ml) liters (l) cubic centimeter (cm^3) cubic meter (m^3)

136) US measures: Units of Length

12 in = 1 ft	3ft = 1 yd	1760 yd = 1 mi	5280 ft = 1 mi

Example problems	Relationships from table above	Multiply with the right conversions so that the units cancel out
1) Convert 20 feet. into inches	1ft = 12 in	$20\ \cancel{ft} \times \dfrac{12\ in}{1\ \cancel{ft.}} = 240\ in$
2) Covert 5280 yards into miles.	1760 yd = 1mi	$5280\ \cancel{yd} \times \dfrac{1\ mi}{1760\ \cancel{yd}} = \dfrac{5280\ mi}{1760} = 3\ mi$
2) Covert 144 inches into yards	1 2 in = 1ft 3in = 1yd	$144\ \cancel{in} \times \dfrac{1\cancel{ft.}}{12\ \cancel{in}} \times \dfrac{1\ yd}{3\ \cancel{ft}} = 4\ yds$

The Problem: Convert	The Relationships You need	Multiply with the right conversions so that the same units cancel out
1) 15 ft. into in.	12 in = 1 ft.	
2) 144 inches into feet.	12 in = 1ft	
3) 72 ft. into yd.		
4) 9 ft. into yd.		
5) 3.5 mi into ft.	1 m = 39.4 in	
6) 2640 ft. into mi.		
7) 120 in into yd.		
8) 3 mi into feet		
9) 8800 ft. into miles		
10) Safiya walked for 2 miles to the school. Express that into feet		

1 km = 1000 m	**1 m** = 100 cm	**1 m = 1000 mm**	**1 cm** = 10 mm

Example problems	Relationships from table above	Multiply with the right conversions so that the units cancel out
1) *Convert 5 km. into meters*	1km = 1000 m	$5 \; km \times \dfrac{1000 \; m}{1 \; km} = 5000 \; m$
2) *Covert 5200 mm into meters.*	1000 mm = 1m	$5200 \; mm \times \dfrac{1 \; m}{1000 \; mm} = 5.2 \; m$
2) *Covert 24000 cm into km.*	100 cm = 1m 1000m = 1km	$24000 \; cm \times \dfrac{1 \; m}{100 \; cm} \times \dfrac{1 \; km}{1000 \; m}$ $= 0.24 \; km$

The Problem: Convert	*The Relationships You need*	*Multiply with the right conversions so that the same units cancel out*
1) 15 m. into cm.	1m = 100 cm.	
2) 1200 mm into cm.		
3) 2.2 m into mm.		
4) 1.2 km into m.		
5) 36000 m into km.	1 km = 1000 m	
6) 1.2 km into m.		
7) 0.25 km into mm.		
8) 360,000 mm into km		
9) 8800 m into km		
10) Vicky walked for 2 km to the school. Express that into meters.		

1 in = 2.5 cm	**1 m** = 39.4 in	**1 mi = 1.61 km**	**1 km** = 0.62 mi

Example problems	Relationships from table above	Multiply with the right conversions so that the units cancel out
3) Convert 5 km. into miles	1km = 0.62 mi	$5\ \cancel{km} \times \dfrac{0.62\ mi}{1\ \cancel{km.}} = 3.1\ mi$
2) Covert 500 cm into inches.	2.5 cm = 1in	$500\ \cancel{cm} \times \dfrac{1\ in}{2.5\ \cancel{cm}} = 200\ in$
4) Covert 64000 in into km.	1m = 39.4 in 1000m = 1km	$64000\ \cancel{in} \times \dfrac{1\cancel{m.}}{39.4\ \cancel{in}} \times \dfrac{1\ km}{1000\ \cancel{m}} = = 1.62\ km$

The Problem	The Relationships You need	Multiply with the right conversions so that the same units cancel out
1) 5 m into inches.	1m = 39.4 in.	
2) 150 inches into meters.	1m = 39.4 in.	
3) 6.2 inches into cm.		
4) 25 cm into inches.	2.5 cm = 1in	
5) 640 miles into km.	1 km = 0.62 mi	
6) 7.2 km into miles.		
7) 0.25 m into inches.		
8) 860 inches into meters	1 in = 2.5 cm 1m =100 cm	
9) 25 m into inches		
10) Guled walked for 2 miles to home. Express that into meters.	1 mi = 1.61 km 1 km = 1000 m	

139) Units of Weight

Now that you are familiar with how to build and use the conversion systems, you have to be able to apply it quickly in converting between units of weight.

A) The U.S Measures	B) The Metric measures	C) Conversion between U.S and Metric measures
1lb = 1 6 oz **1T = 2000 lb**	1 kg = 1000 g 1 g = 1000 mg	1 lb = 454 g 1 kg = 2.2 lb 1 ton = 907.2 kg

The Problem: Convert	The Relationships You need	Multiply with the right conversions so that the units cancel out!
1) 12 pounds into ounces	1 lb = 16 oz	$12\text{lb} \times \dfrac{16 \text{ oz}}{1\text{lb}} = 192 \text{ oz}$
2) 0.2 Kg to ounces.	*1kg = 2.2 lb* *1 lb =16 oz*	$0.2 \text{ kg} \times \dfrac{2.2 \text{ lb}}{1 \text{ kg}} \times \dfrac{16 \text{ oz}}{1\text{lb}} = 7.04 \text{ oz}$
3) 4000 g to kg.	1 kg = 1000 g	
4) 22 lb into g.		
5) 0.25 T into g		
6) 15 g into milligrams		
7) 20 lb into g		
8) 1 T into kg.		
9) 2400 grams into pounds.		
10) 360 oz into pounds		

140) Units of Capacity or Volume

My friend visited us from Denmark one time. Anytime I buy few gallons of fuel, he can't help but to compare prices in Europe with that of U.S. How many liters are five gallons? After I give my estimate he says, it is cheaper in the U.S. These relationships below helped me for the estimates:

The U.S Measures		The Metric measures	Conversion between U.S and Metric measures
8 fl oz = 1 cup 2 c = 1 pt 2pt = 1 qt	1 c = 0.25 qt 1 gal = 4 qt 1 qt = 32 fl oz	1 L = 1000 ml 1ml = 1 cm^3 1000 l = 1 m^3	1 L = 1.06 qt 1qt = 946 ml 1 c = 0.24 l

The Problem: Convert	The Relationships You need	Multiply with the right conversions so that the units cancel out
1) 8 cups into quart	1 c = 1 pt 2 pt = 1 qt	$8 \text{ c} \times \dfrac{1 \text{ pt}}{1 \text{ c}} \times \dfrac{1 \text{ qt}}{2 \text{ pt}} = 4 \text{ qt}$
2) 2 L into gallons.	1L = 1.06 qt. 4 qt = 1 gallon =	$2 \text{ L} \times \dfrac{1.06 \text{ qt}}{1 \text{ L}} \times \dfrac{1 \text{ gal}}{4 \text{ qt}} = 0.8 \text{ gal}$
3) 4000 ml into L.		
4) 2 gal into pt.		
5) 25 c into L		
6) 10 L into fl oz		
7) 4 pints into milliliters		
8) 20 gal into ml		
9) 473 ml into fl oz	1qt = 946 ml 1 qt = 32 fl oz	
10) 1000 cups into ml.		

141) Add and Subtract Mixed Units

There is an African proverb that says "oil and water cannot be mixed". When adding and subtracting units of measures such as feet and inches, pounds and ounces, make sure that you change them to the same units. Never try to mix oil and water. It won't happen!

Example 1: Add: (a) 12 lb 15 oz. + 2lb 6 oz.   (b) 6ft 8 in + 3 ft 10 in

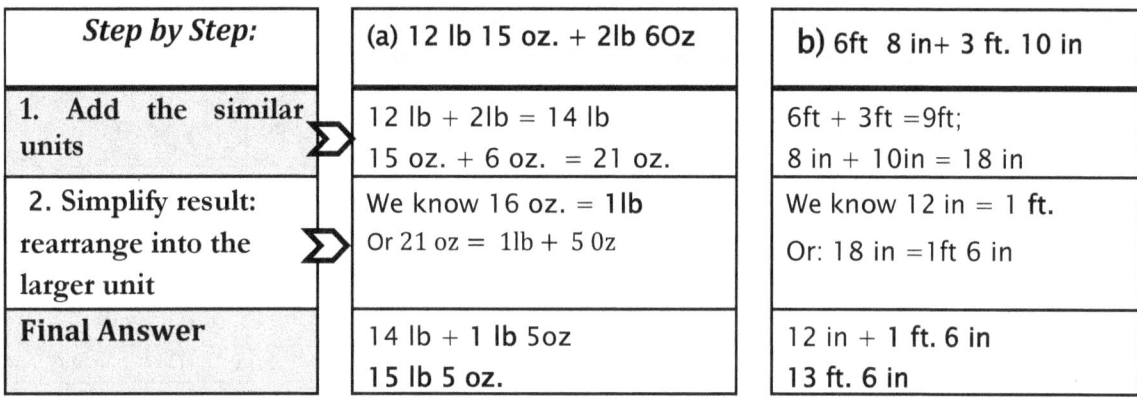

Step by Step:	(a) 12 lb 15 oz. + 2lb 6Oz	b) 6ft 8 in+ 3 ft. 10 in
1. Add the similar units	12 lb + 2lb = 14 lb 15 oz. + 6 oz. = 21 oz.	6ft + 3ft =9ft; 8 in + 10in = 18 in
2. Simplify result: rearrange into the larger unit	We know 16 oz. = 1lb Or 21 oz = 1lb + 5 0z	We know 12 in = 1 ft. Or: 18 in =1ft 6 in
Final Answer	14 lb + 1 lb 5oz 15 lb 5 oz.	12 in + 1 ft. 6 in 13 ft. 6 in

Example 2: Subtract (a) 12 ft 5in − 7ft 3in (b) 12 ft 5in − 7ft 9in

Subtractions could be either straight forward (Example 2a) or require borrowing (Example 2b)

Step by Step:	(a) 12 ft 5 in - 7ft 3 in	(b) 12 ft 5in − 7ft 9in
✓ Subtract similar units. ✓ Borrow if you need to	$12ft − 7\,ft = 5ft$ $5\,in − 3in = 2\,in$ Final answer: <u>5ft 2in</u>	We can't subtract 9 from 5. So, borrow 1ft from the 12 ft and change it to inches 11 ft 17 in − 7ft 9 in = <u>4 ft 8 in</u>

142) Units: Find the sum or difference:

	a)	b)	c)
1)	9 ft 4 in. + 7 ft 9 in	10 *yd* 5*ft* − 7yd 8ft	7 *m* 70 *cm* + 5m 40 cm
2)	9 m 20 cm − 7 m 30 cm	1 gal 2 qt. + 7 gal 4qt.	12 kg. 200g − 8 kg. 700g
3)	13g 200 mg − 4 g 900 mg	8c 2fl. oz. − 2c 5 fl oz.	4 ft 1 in. − 2 ft 9 in
4)	15 L 500 ml + 7L 600 ml	9 km 400 m − 7 km 800 m	3 *lb* 2 oz. − 2 *lb* m 3 oz
5)	8 ft 2 in. − 2 ft 10 in	8 *yd* 1*ft* − 4yd 2 ft	10 *m* 10 *cm* − 5m 40 cm

APPENDIX 1: ANSWERS OF SELECTED EXERCISE-1

1) Primes	1) 3, 5, 7, 11, 13,17 3) 41, 43, 47, 53 5) 67, 71,73,79
2) Prime Factor	1) 2^4 3) $2^3 \times 7$ 5) 3×13 7) $2^2 \times 5^2$ 9) 3×5^2 11) 2^7
3) Factoring:	1) 1, 2, 4, 8, 16, 3) 1, 2, 4, 7, 8, 14, 28, 56, 5) 1, 2, 3, 4, 6, 7, 12, 14, 21, 28, 42, 84 **7)** 1, 2, 4, 8, 17, 34, 68, 136 **11)** 1, 2, 4, 5, 10, 20, 25, 50, 100
4) GCF1	1) 2 3) 7 5) 11 7) 1 9) 10
5) GCF2	1) 7 3) 9 5) 2 7) 8 9) 6
6) LCM1	3) 28 5) 22 7) 144 9) 20
7) LCM2	1) 14 3) 18 5) 56 7) 80 9) 72
8) Improper & Mixed	**1)** $1\frac{3}{4}$ 3) $5\frac{2}{3}$ **5)** $3\frac{3}{8}$ 7) $6\frac{1}{2}$ **9)** $12\frac{3}{6}$ 11) $3\frac{5}{6}$ **13)** $12\frac{3}{7}$ 15) $6\frac{3}{5}$ **17)** $\frac{16}{13}$ 19) $\frac{13}{2}$
9) Simplify	**1)** $\frac{1}{4}$ 3) $\frac{1}{2}$ **5)** $\frac{1}{3}$ 7) $\frac{1}{3}$ **9)** $\frac{1}{4}$ 11) $\frac{5}{8}$ **13)** $\frac{4}{5}$ 15) $\frac{3}{4}$ **17)** $7\frac{3}{5}$ 19) $5\frac{2}{3}$ **21)** $4\frac{1}{5}$
10) Add/sub Fractions	**1)** $\frac{3}{5}$ 3) $\frac{1}{2}$ **5)** $\frac{1}{2}$ 7) $4\frac{3}{5}$ **9)** $9\frac{3}{10}$ 11) $\frac{1}{3}$ **13)** $\frac{9}{10}$ 17) $1\frac{1}{20}$ 19) $3\frac{3}{8}$
11) Unlike Fractions	**1)** $1\frac{3}{8}$ 3) $\frac{13}{20}$ **5)** $\frac{1}{6}$ 7) $\frac{11}{12}$ **9)** $\frac{29}{30}$ 11) $3\frac{3}{4}$ **13)** $7\frac{8}{15}$ 17) $1\frac{13}{20}$ 19) $4\frac{17}{84}$
12) Multiply Fractions	**1)** $\frac{2}{25}$ 3) $\frac{3}{8}$ **5)** $\frac{1}{15}$ 7) $\frac{1}{12}$ **9)** $\frac{21}{32}$ 11) $1\frac{1}{2}$ **13)** 14 15) 2 17) 15
13) Divide Fractions	**1)** $\frac{2}{25}$ 3) $1\frac{1}{3}$ **5)** $\frac{1}{5}$ 7) 1 **9)** $1\frac{8}{27}$ 11) $1\frac{1}{6}$ **13)** $3\frac{3}{5}$ 15) $\frac{24}{65}$ 17) $\frac{68}{81}$
14) What is Percent?	**1)** 25% 3) 10% **5)** Playing & Reading 7) 75%
15) Fraction into Percent	**1)** 75% 3) 25% 5) 20% 7) 83.3% 9) 8.3% **11)** 41.7% 13) 4.2 15) 1.2%
16) Mixed to Percent	**1)** 350% 3) 725% 5) 920% 7) 780% 9) 1125% **11)** 1233%
17) Mixed to Percent-2	**1)** 350% 3) 725% 5) 875% 7) 540% 9) 780% **11)** 1125% 13) 205% 15) 1701%
18) Percent to fractions	**1)** $\frac{13}{100}$ 3) $\frac{17}{100}$ **5)** 2 7) $\frac{6}{25}$ **9)** $\frac{55}{100}$ 11) $\frac{21}{50}$ **13)** $7\frac{1}{25}$ 15) $1\frac{7}{25}$
19) Mixed Percent to Fraction	**1)** $\frac{9}{400}$ 3) $\frac{21}{400}$ **5)** $\frac{2}{125}$ 7) $\frac{33}{400}$ **9)** $\frac{31}{350}$ 11) $\frac{4}{125}$

Answers- Continued-2

20) Decimals to Percent	1) 130% 3) 102% 5) 650% 7) 705% 9) 950%
21) **Percent to decimals**	1) 0.13 3) 0.40 5) 0.013 7) 2.0 9) 0.045 11) 0.40 13) 0.055
22) **decimal % to Fraction**	1) $\frac{13}{1000}$ 3) $\frac{17}{1000}$ 5) $\frac{1}{50}$ 7) $\frac{11}{5000}$ 9) $\frac{11}{200}$ 11) $\frac{9}{200}$
23) **Mixed Review**	3) $\frac{7}{10}$, 70% **5) 0.75, 75% 7)** $\frac{3}{200}$, 0.015 **9)** $5\frac{2}{5}$, 540% 11) $\frac{69}{500}$, 0.138
24) **Percent: Find part**	1) 4 3) 75 5) 41 7) 168 9) 50
25) **Finding the Percent**	1) 4% 3) 25% 5) 4.76% 7) 2.5% 9) 2%
26) **Finding the base**	1) 300 3) 300 5) 500 7) 1000 9) 470
27) **Percent Mix Practice**	1) 10 3) 4% 5) 12.5% 7)40% 9) 80%
28) **Percent Change**	1) 20% 3) 16% 5) 75% 7)92.11% 8) 33%
29) **Tax, Discount etc.**	1) 480% 3) 2750 5) 336 7)133.76 8) 14.16 hours.
30) **Perfect Squares**	The squares of 11, 12, 13, 14 and 15 are 121, 144, 169, 196 and 225.
31) **Square Roots**	The square roots of 225, 289, 324, 361, are 15, 17, 18, and 19.
32) **Cubic, quartic Roots**	1) 1, 3) 3, **5)** 1, 7) 0, **9)** 4, 11) 4, 13) 15, **15)** 5.
34) **Order of Operations**	1) 20, 3) 14, **5)** 25, 7) 6, **9)** 32, 11) 20
35) **Advanced Order of Op.**	1) 11, 3) -4, **5)** -1, 7) 198, **9)**0
37) **Real numbers**	3) R **5)** R, I, W 7) R, I, W, N **9)** R, I 11) IR, 13) R 15) R, I ,N, W
38) **Algebraic Properties I**	1, 3, 5, 7, 8 10 are commut. 2 4, 9 are distributive, 6, 11, and 12 are associ.
39) **Algebraic Properties II**	1,2, 5,9,12 iden prop, 3 & 7 are add inverse, 4, 8 mult. Inverse, 6 is zero p.
40) **Integers: Absolute Val.**	1,2, 4, and 7 are negative. 3, 5 & 6 are positive, 9) 12 11) 17 13) 22 15) 45
41) **Ordering Integers**	**A)** 1) > 3) > 5) < 7) > 9) < 11) > **B) 1) -9,-7,-5,0, 5,** I-10 I
42) **Add Integers**	**1)** -10, **3)** 4, **5)** -9, **7)** 6, 9) 4 11) -5 13) -37 15) -21 17) -24 19) 5
43) **Subtract Integers**	**1)** -4, **3)** 12, **5)** -17, **7)** 6 , 9) 12, 11) 23 , 13) -19, 15) -13 17) -6, 19) -75
44) **Algebraic expressions**	**1)** -10x, **3)** 8x+8y, **5)** -2a-4b, **7)** -17m , 9) -8x-9, 11) 5y , 13) -17t, 15) -21t
45) **Evaluate Algebra Exp 1**	**1) -4, 3)** 6, **5)** -10, **7)** -18, 9) 7, 11) 15, 13) -36, 15) -16, 17) 14, 19) -21
46) **Multiply Integers**	**1)** 12, **3)** -32, **5)** -52, **7)** 30, 9) 4, **11)** -1, **13)** 12, **15)** 28, **17)** -320, **19)** -10
47) **Divide Integers**	**1)** 2, **3)** -2, **5)** 3, **7)** -6, **9)** 1, **11)** -9, **13)** -5, **15)** 9, **17)** 2, **19)** -25
48) **Algera Multipl & Divi**	**1)** 24, **3)** 12yz, **5)** 8ab, **7)** 24xy, **9) 3xy, 11)** 24x, **13)** 0, **15)** 30x, **17)** 16w, **19) 12w**
49) **Evaluate Algebra Exp2**	**1)** 96, **3)** 72, **5)** -6, **7)** 48, **9) -48, 11)** 12, **13)** 32, **15)** 216 **17)**-32, **19) 60**
50) **Evaluate Algebra Exp3**	**1)** 14 **3)** 72, **5)** 18, **7)** 9, **9)** 6, **11)** 36, **13)** 4, **15)**14 **17a)** 2+2x **17b)** 14
51) **One Step Equations**	**1)** 20 , **3)** 22, **5)** 7, 7) 15, **9)** 1, **11)** 5 , **13)** 12, **15)**3

Answers- Continued-3

52) Solve Equations: Addition	1) 6 , **3) -3, 5)** 7, **7)**-6, **9) 0, 11)**-9 , **13)** 4, **15)**-20, 17) 10, 19) 9
53) Solve Equations: Subtract	1) 14 , **3) 2, 5)** 15, **7)**-4, **9) 22, 11) 7 , 13)** -4, **15)** 6, 17) 20, 19) -9
54) Solve Equations (+ /-)	1) -4 , **3) -10, 5)** 7, **7)**-4, **9) -12, 11) 7 , 13)** -4, **15)** -20, 17) 4, 19) 1
55) Solve Equation: Multiply	1) 5 , **3) -5, 5)** 6, **7)** 8, **9) 3.5, 11) -2 , 13)** -2, **15)** 1, 17) -9, 19) -0.5
56) Solve Equation: Division	1) 10 , **3) -96, 5)** 36, **7)** 36, **9) 14, 11) 16 , 13)** -32, **15)** -120, 17) 10, 19) 0
57) Solve Equation: (× & ÷)	1) 4 , **3) 14, 5) -9, 7)** 1, **9) 55, 11) 0 , 13) -16, 15)** -5, 17) 51, 19) -6
58) Solve Equation: (Fraction)	1) $\frac{1}{5}$, **3)**$\frac{4}{7}$, **5)** $\frac{4}{3}$, **7)**-$\frac{7}{40}$, **9)** $4\frac{3}{7}$, **11)** $-\frac{1}{4}$
59) Solve Equation: (Decimals)	1) 0.8 , **3)**-1.4, **5)**0.5 , **7)**-10, , **9)**-12, **11)** -0.7 , **13)** -0.1, **15)** -500, 17)-17
60) Solve Equation: (Review)	1)-7 , **3) 18, 5)**-96 , **7)** -0.2, , **9)**-65, **11) 0 , 13)** -24, **15)** $\frac{1}{3}$, , 17)-22
61) Math Language: Translate	1) 15 , **3) 40, 5) 15/7 , 7)** 1,
62) Math Language: Variables	**1) x+11 3) 20+x 5) 5x-15 7) 7/x 9) 9+6 11) 2x+2x**
63) Equations: Word problems	**1) X+3 =10; 7 3)** x+y =27; 13 **5)** x-10=42, 52 **7)** 12/x =3; 4 **9)** 7x=63; 9
64) Two Step Equations:	**1) 7 , 3) 21, 5) -45, 7) 1 , 9) -70, 11) 24, 13) -3 , 15) -15**
65) The Distributive Property	**1) -6x+24 , 3) -3z+12y 5) -4a+8b, 7)**-6x+12y **9) -3x +3y 11) 15-12x**
66) Combining Like Terms	**1) 9x-21 , 3) -4y 5) -5, 7)** 21x+6 **9)** x −y, **11) -4, 13) 2x-10, 15) 4y**
67) Algebraic Fractions	**1) 2x 3)** $\frac{7}{9}x$ **5)** $\frac{6}{35}y$ **7)**$\frac{10}{7}t$ **9)** $\frac{9}{10}t$ **11)** $\frac{-17}{12}m$ **13)**$\frac{2}{3}t$ **15) 0**
68). Variables on Both Sides I	**1) 2 3) -3 5) -4 7)** $\frac{1}{2}$ **9)**−1 **11) -1 13) -3 15) -1 17) 1**
69) Multistep Equa: Review	**1)** $\frac{16}{3}$ **3)** $\frac{20}{3}$ **5) 10 7) 1 9) 1 11)** $\frac{3}{5}$ **13) 0.0125 15) -7**
70) Graph Inequality	
71) write Inequality	**3)** $x \leq 5$ **5)** $-2 < x \leq 0$ **7)** $x \leq 10$ **9)** $-3 < x < 3$

,

Answers- Continued-4

72) Exponents: The Basics	1) $13x^2$ 3) $2x^3 + 6$ 5) $7y^2 + 2x^2$
73) Add/Subtract Exponents	1) 1 3) 124 5) 19 7) $9y^7$ 9) 247 11) $3y^4$ 13) $-7y^4$ 15) $4y^3 + 19y^2$
74) Multiply/Divide Exponents	1) 3^8 3) 49 5) 1728 7) y^4 9) 15 11) 0
75) Power of Power	1) x^9 3) y^6 5) y^2 7) y^2 9) x^{15} 11) 81
76) Negative Exponents	1) $\frac{1}{x^3}$ 3) $\frac{1}{y^6}$ 5) 1 7) $\frac{1}{x^3}$ 9) $\frac{1}{x^5}$ 11) 1
77) Evaluate Exponents	1) 81 3) 64 5) 256 7) 12 9) 728 10) 50
78) Scientific Notation-1	1) 35200 3) 0.000561 5) 8020 7) 0.46 9) 650000 11) 8213000
79) Scientific Notation-2	1) 4.5×10^{-4} 3) 4.1×10^4 5) 2.1×10^{-2} 7) 6.1×10^5 9) 3.2×10^{-3}
80) Eval. Roots /Radicals	1) 3 3) 7 5) 4 7) 9 9) 44 11) 40 13) 3 15) $6\sqrt{5}$
82) Write Ratios: simplify	1) $\frac{4}{5}, \frac{5}{8}$ 3) $\frac{7}{13}$ 7 5) $\frac{412}{147}$ 7) $\frac{5}{14}$
84) Solve Proportions	1) 14 3) 0.3 5) $\frac{50}{3}$ 7) $\frac{9}{2}$ 9) $\frac{18}{13}$ 11) 5.53 13) 0.05 15) 0.7
86) Proportion Problems:	1) 13 3) 137.5 5) 24 7) 4 cups
88) Graph: Ordered Pairs	A (2,3), B(-2,-2), D (5,-1), F (-3,4) , I(-2,3), K (-3,-4), N (3,-3)

90) Find Ordered Pairs

3) x -2y =6				5) 2x -y =4				9) 2x +y =-2		
x	y	(x, y)		x	y	(x, y)		x	y	(x, y)
-2	-4	(-2,-4)		-2	-8	(-2,-8)		0	8	(0, 8)
0	-3	(0,-3)		0	-4	(0,-4)		1	6	(1, 6)
1	-2.5	(1,-2.5)		1	-2	(1,-2)		2	4	(2, 4)
2	-2	(2,-2)		2	0	(2,0)		3	2	(3, 2)

92) Graphing:

1) y=2x	3) y=-4x-2	5) y=3x+1

94) Find Slope from Graph	1) slope= -1 3) slope =2 5) slope =-1

95) Slope from Graph	1) -1, $-\frac{2}{3}$ 3) 1, $-\frac{1}{9}$ 5) $\frac{2}{7}$, $-\frac{3}{2}$ 7) $-\frac{9}{7}$, $-\frac{1}{5}$ 9) 1, -2

96) Equation of Line 1	A) 1) y= 2x+1 3) y=-4x-2 5) y =5x 7) y=-5 13) -5,7 15) -6,-4 17) 1,0

97) $y = 4x - 2$	$y = 2x + 2$	$y = 3x$

98) Slope/ Inter	**1)** m=2; y-inter =3 **3)** m=-4 ; y-inter =-2 5) m=1, y-intercept =0
100) Read Graophs	1) Friday 3)Wednesday & Tuesday 5) 600 7) year 1 9) 30 and 35
102)Name Angles	1) <AOE or <EOA 3) <DOC or COD 5) <BAC, <CAB or <A 7) <IGH, <HGI or <G
103) Measure Angles	1) 45° 3) 135° 5) 135° 7) 45°
104) Classify Angles	**1) right 2) Obtuse 3) acute 4) x=41 5) y=135 6) z=92**
106) Angles	1) 2&3, 1&4, 5&8 3) 1&2, 3&4, 5&6 5) 1&7, 2&8. 7) <1 &<4. 9)73°

Answers- Continued-6

107) Triangles	1) isocels 3) scalene 5) scalene 7) $100°$ **9)** $50°$
108) Pythagorean	1)10 3) 6.71 5)13.23 7) 8.06
109) Distance Formulla	1) 2.83 3) 5 5) 5 7) 12.2 9) 18.87
110) Special Triangles	1) x= $4\sqrt{3}$, y =8 3)$x = 16, y = 4$ 5) $x = 5\sqrt{2}$; y =5; 7) y=24, $x = 12\sqrt{3}$
113) Area and Perimeter	1) 9 in^2; 12 in 3) 34 in^2;26 in 5) 128 in^2;32 in 7) 36 cm^2;28 cm 9)216cm^2; 64cm 11) Area of triangle= 3 in^2; Area of Trapezoid= 21 in^2
115) The Circle	1) C= 12.56 cm; A = 12.56 cm^2 3) C= 43.96 cm; 153.86 cm^2 5) Circle A; d=8, C = 25.12 cm , A = 50.24cm^2 7) False 9)
116) Geometry Review	1) False 3) False 5) False 7) True 9) True 11) 24 ft^2 13) 45cm^2 15) 3
117) Cube	1) 150 cm^2 3) 600 cm^2 5) 864 cm^2 7) 726 cm^2, 1331 cm^3 9)486 cm^2 , 729 cm^3
118) Rectangualar Prism	1) 148 ft^2 3) 112 ft^2 5) 148 120 ft^3 7) 324 cm^3
119) Sphere	1) 113 cm^2 3) 1256 cm^2 5) 1519.8 cm^2 7) 11488 cm^3 9)14,130 cm^3
120) Cylinder	1) 207 m^2 3)1884 m^2 5) 2211 m^2 7) 414 m^3 9) 2411m^3
121) Cone	1) 104 ft^2 3) 393 ft^2 5) 509ft^2 7) 33.5ft^3 9) 804ft^3
122) Pyramid	1) 96 cm^2 **3) 88** cm^2 **5) 24** cm^3 **6) 43** cm^3
123) Solid Geo Review	1) 1350 cm^2 **3) 1520** cm^2 **5) 754** cm^2 **7) 10125** cm^3 **9) 4187** cm^3
125) Statistics 1	1) mean = 89, med =90, mode=95, Range= 15 3) 100
126) The 5 Number	Minimum: 135, Quartile Q1: 145, Median: 153.5, Quartile Q3: 175 Maximum: 210
128) Probability 1	1) $\frac{1}{6}$ 3) 0 5)$\frac{2}{3}$ 7)$\frac{5}{6}$ 9)$\frac{1}{3}$ 11)$\frac{1}{4}$ 13)$\frac{1}{2}$
129) Probab: Compound	1)$\frac{1}{16} = 0.0625$ 3)$\frac{1}{216} = 0.0046$
130) Probab: Either/ OR	a)$\frac{5}{12} = 0.4167$ b) $\frac{2}{3} = 0.6667$ c) $\frac{12}{12} = 1$
131) Probab: Either/ OR 2	1) a)$\frac{2}{52} = \frac{1}{26}$ b)$\frac{4}{52} + \frac{26}{52} - \frac{2}{52} = \frac{28}{52} = \frac{7}{13}$ 2) a)$\frac{2}{52} = \frac{1}{26}$ b) $\frac{2}{52} + \frac{13}{52} - \frac{2}{52} = \frac{13}{52} = \frac{1}{4}$
132) Probab: And Events	$\frac{3}{4} \times \frac{1}{4} = \frac{3}{16}$

Answers Continued 7:

133) Probab: And Events 2	a) $\frac{4}{52} \times \frac{4}{51} = \frac{4}{663}$ b) $\frac{4}{52} \times \frac{3}{51} = \frac{1}{221}$ c) $\frac{2}{52} \times \frac{1}{51} = \frac{1}{1326}$
134) Mixed Probability	1) $\frac{2}{6} = \frac{1}{3}$ 3) $\frac{3}{6} \times \frac{3}{6} \times \frac{3}{6} = \frac{1}{8}$ 5) a) 0 b) $\frac{1}{5}$
136) Measurements: Length	1) 180 in 3) 24 yd. 5) 18,480 ft. 7)3 yd 1ft. 9) 1.67 mi
137) Metric System: Length	1) 1500 cm. 3) 2200 mm 5) 36 km 7) 250 mm 9) 8.8 km
138) Metric & U.S measures	1) 196.9 in 3) 15.7 cm 5) 396.8 mi 7) 9.9 in 9)985 in
139) Units of Weight	3) 4 kg 5) 226796 g 7) 9080 g 9) 5.3 lb
140) Units of Volume	3) 4 l 5) 6 l 7) 1892 ml 9) 16 ml.
142) Units: Add & Subtract	1a) 17ft 1 in 1b) 2 yd. 1c) 13m. 10cm. 3a) 8g, 300mg. 3b) 5c 5fl.oz. 3c) 1ft 4 in 5a) 5ft 4in 5b) 3 yd 2ft 5c) 4m 70cm

Congratualtios for your determination and effort. Keep going and start higher algebra!

www.ingramcontent.com/pod-product-compliance
Lightning Source LLC
Chambersburg PA
CBHW081724170526
45167CB00009B/3695